MW00512991

CATALYTIC CRACKING

CHEMICAL INDUSTRIES

A Series of Reference Books and Textbooks

Consulting Editor
HEINZ HEINEMANN
Heinz Heinemann, Inc.,
Berkeley, California

CATALYTIC CRACKING
Catalysts, Chemistry, and Kinetics

Bohdan W. Wojciechowski
Queen's University
Kingston, Ontario, Canada

Avelino Corma
Instituto de Catalisis y
Petroleoquimica C.S.I.C.
Madrid, Spain

Marcel Dekker, Inc.   New York • Basel

ISBN 0-8247-7503-8

Marcel Dekker, Inc.
270 Madison Avenue, New York, New York   10016

Current printing (last digit):
10 9 8 7 6 5 4 3 2 1

Printed in the United States of America

To Brisa and Margot. And to all our colleagues, past, present, and future; Their work advances the understanding of catalytic cracking and contributes to the fundamental comprehension of catalysis.

PREFACE

Scientific research is often compared to exploration. Expeditions (research programs) are mounted, discoveries made, and reports presented. As the information from the various expeditions accumulates, the time finally comes when a map, however imperfect, may be drawn. Curiosuly, few such maps of catalytic cracking have been attempted.

This book is our attempt at such a map. We do not pretend to have the final version nor can we answer for any erroneous reports from the various expeditions. We can only set down what we believe to be reliable, and annotate the text with comments equivalent to "here there be demons" on the charts of old.

We offer this work as a guide for future work and commentary, so that the field of catalytic cracking can be systematized and explored in a rational manner. Each new discovery can then be placed in relation to previously explored terrain, and thereby fill a gap on future maps.

We have also presented new methodologies for the exploration of this field. These, we hope, will remove the ambiguities so often present in previous reports. By raising our expectations regarding future reports, we can substantially reduce the time and effort that will be required to improve the mapping.

This volume is designed so that a complete reading from beginning to end is recommended, rather than a sampling of the "good parts." We feel that a careful reading from beginning to end will answer many questions that might persist in the mind of the reader who chooses to read selectively. For those whose curiosity is still not satisfied by a thorough reading, there is a rich selection of references, many of them reviews.

We would like to thank Amelia Laing, science librarian at Queen's University, for her assistance with the editing and verification of the references and Margot Wojciechowski for her many helpful comments and editorial assistance during the preparation of this book. Their help was beyond price.

Bohdan W. Wojciechowski
Avelino Corma

CONTENTS

1

Introduction

Since its introduction some 40 years ago, catalytic cracking has become one of the most important of the petroleum refining processes. The reason for this durability and the most significant characteristic of the cracking process is its flexibility in treating the variety of feedstocks available from whatever crudes may need to be refined. This flexibility becomes increasingly important as refineries are obliged to resort to heavier crudes containing refractory or poisonous constituents, because of shortages and the high price of the more desirable feedstocks.

Following its introduction by Houdry and Joseph [1], catalytic cracking has been the subject of many investigations designed to clarify the fundamental chemical processes involved and the kinetics of the overall reaction. Most such investigations have been conducted by petroleum companies and held confidential to protect patents and know-how. This fact speaks loudly for the value of such information, although it also suggests that the majority of work is very "applied." It is only in the last 15 years or so that some of the industrial results have been published. Because of this lack of communication, and because of the highly applied nature of the work in the field, fundamental understanding of the various phenomena involved is far from complete.

The lack of understanding has been further complicated by the fact that the principal property of cracking catalysts, and the reason for their success, is their ability to deal with the complex feeds that are used commercially, by means of the large number of reactions that occur, both in parallel and consecutively. Such complexity makes the establishment of the reaction network and the development of sound kinetic laws very difficult

To begin to understand the effect of cracking catalysts on the feed stock, we can look at the product distribution resulting from thermal cracking and compare this to the distribution obtained by catalytic cracking [2]. Such results are shown in qualitative terms in Table 1.1. There we see that the differences between the product distributions are such that the two cracking processes must surely proceed by entirely different mechanisms.

Table 1.1   Comparison of Products Obtained by Catalytic and Thermal Cracking

Hydrocarbon cracked	Thermal cracking products	Catalytic cracking products
n-Hexadecane	C_2 is the major product C_1 produced in large amounts C_4 to C_{15} olefins in products No branched-chain products	C_3 to C_6 are the major products No olefins larger than C_4 Branched-chain paraffins present in products
Aliphatics	Little aromatization at $500°C$	Significant aromatization at $500°C$
Alkyl aromatics	Cracking occurs within the side chain	Dealkylation is the dominant cracking reaction
n-Olefins	Slow double-bond isomerization Little skeletal isomerization	Rapid isomerization of double bond Rapid skeletal isomerization
Naphthenes	Cracking is slower than that of paraffins	Cracking rates comparable to those of paraffins

Source: Ref. 2

In the case of thermal cracking, it is well established that cracking occurs via the intervention of free radicals in a variety of complicated chain mechanisms. Since the product distribution is so different for the catalytic cracking reaction, one has to imagine that the latter involves a non-free-radical mechanism.

In 1943, Grosse [3] proposed that the physical adsorption of hydrocarbons on a catalyst surface facilitates thermal cracking by lowering the activation energy of the process. This was no more than a specific application of the common view of catalytic action current at the time and does not provide the necessary departure from gas-phase mechanisms to explain the differences in product distribution.

Taking a different tack, Turkevich and Smith [4] suggested that cracking takes place via "hydrogen exchange" in which the hydrocarbon adsorbs on specific locations in certain regions of the catalyst surface. Their theory requires that the catalyst have the capacity to give up or accept an atom of hydrogen from a distance of 3.5 Å, which is the distance between atoms of hydrogen on a primary and a tertiary carbon in an aliphatic chain. Although this and other such hypotheses have led to fruitful ideas, they cannot explain many of the phenomena actually observed in catalytic cracking.

The first durable idea on the mechanism of catalytic cracking was that of Gayer [5], who showed in 1933 that a catalyst consisting of alumina supported on silica possesses acidic properties. This observation, confirmed later by others, led to speculation that these acid sites may in fact be the catalytically "active centers." About the same time, Whitmore [6] proposed acid sites as the active centers in his theory of carbenium ion reactions. It is a pity that this proposal, first put forth in 1934, was not followed up until the late 1940s.

In 1947, Hansford [7] made the first detailed attempt to explain the mechanism of catalytic cracking in terms of ionic reactions. In his original work Hansford described the mechanism in terms of carbenium (then called carbonium) and carbene ions, but later he modified the mechanism to include only carbenium ions [8,9].

In 1949, Thomas [10] independently proposed a mechanism similar to that of Hansford. The carbenium ion postulate is now known to agree with the observed facts and to explain the following specific characteristics of catalytic cracking:

1. The mechanism of the carbon-carbon bond rupture

2. The greater susceptibility to cracking of compounds containing a tertiary hydrogen

3. The higher cracking rates of C_6 and higher paraffins

4. The high rates of dealkylation of chains containing three or more carbons in alkyl aromatics

5. The preferential formation of products of three or more carbons from linear hydrocarbons

6. Skeletal isomerization

7. Hydrogen transfer

8. Selective saturation of tertiary olefins

As a result it is generally accepted today that the initial event in a cracking reaction is the formation of a carbocation. Thus a better knowledge of the nature of carbocations, how they are formed, and how they evolve should lead to a better understanding of the reactions taking place during cracking.

REFERENCES

1. E. Houdry and A. Joseph, *Bull. Assoc. Fr. Tech. Pet.*, *117*: 177 (1956)

2. P. B. Venuto and E. T. Habib, Jr., *Fluid Catalytic Cracking with Zeolite Catalysts*, New York, Marcel Dekker, 1979

3. A. V. Grosse, *Ind. Eng. Chem.*, *35*: 762 (1943)

4. J. Turkevich and R. K. Smith, *J. Chem. Phys.*, *16*: 465 (1947)

5. F. H. Gayer, *Ind. Eng. Chem.*, *25*: 1122 (1933)

6. F. C. Whitmore, *Ind. Eng. Chem.*, *26*: 94 (1934)

7. R. C. Hansford, *Ind. Eng. Chem.*, *39*: 849 (1947)

8. R. C. Hansford, P. G. Waldo, L. C. Drake, and R. E. Honig, *Ind. Eng. Chem.*, *44*: 1108 (1952)

9. R. C. Hansford and C. Rowland, *Adv. Chem. Ser.* (Heterogeneous Catalysis), *222*: 247 (1983)

10. C. L. Thomas, *Ind. Eng. Chem.*, *41*: 2564 (1949)

2

Fundamentals of Carbocation Behavior

2.1   INTRODUCTION

Carbocations can be thought of as organic cations with the charge
more or less associated with a carbon atom [1]. The idea that
such organic cations exist was proposed 70 years ago by workers
studying triphenylmethane derivatives in acid solutions [2].
Some eight years later it was suggested that carbenium ions may
act as intermediates in certain organic reactions [3]. This was
followed by the classical proposals of Whitmore [4] and his school,
who developed a cohesive picture of carbenium ion reactions in
the liquid phase. An important aspect in all such considerations
is the ability to predict the type of carbocations that will be
formed, the stability of such ions, and the reaction paths they
might follow.

The existence of carbocations in superacid media has been
directly observed by nuclear magnetic resonance (NMR) and
other techniques. These and other developments are discussed
by Olah et al. in a series of highly pertinent papers [5,6]. In
the following we will accept the existence of such ions and be
concerned with such entities only as they pertain to catalytic
cracking.

2.1.1   Nomenclature, Structure, and Stability
of Carbocations

According to the IUPAC nomenclature, ions that were formerly
called carbonium ions, such as $CH_3^{\oplus}$, are to be called *carbenium
ions*. Other positively charged species of the type $CH_5^{\oplus}$ will be
called *carbonium ions*. The term *carbocation* covers both of the
above and is the general term for positively charged organic
species. Olah [7,8] has also introduced the terms *classical ions*
and *nonclassical ions* to distinguish the trivalent carbenium ions
from the pentavalent carbonium ions.

Carbenium ions are generally planar or nearly planar, as
shown by NMR and Raman studies [9]. If no skeletal rigidity or
steric hindrance prevents planarity, an sp^2 hybridized electron-
deficient center develops. In such structures the interactions of
neighboring groups with the vacant p orbital of the carbenium
center contribute to ion stability via charge delocalization. This
can be due to atoms having unshared electron pairs, to hyperconjugation, or to conjugation with bent bonds or with a electron
system by allylic stabilization [8]. In view of this variety of

effects and the superimposed range of strengths of each of them, there is a large range of carbocation structures, types and reactivities. Other carbocations, such as the alkynium $RCH=C^{\oplus}R$ and the benzynium $(C_6H_5)^{\oplus}$, have a carbocation center with a coordination number of 2.

In contrast to the di- or tricoordinated carbenium ions, carbonium ions are associated with carbon atoms with four or five ligands. The existence of such species has been reported in superacid solvents [5]. Unlike the rather well defined trivalent classical carbocations, nonclassical ions are more loosely defined [10−12]. The simplest penta-coordinated carbonium ion, $CH_5^{\oplus}$ [13−18], is thought to assume several structures, with the symmetries shown in Figure 2.1. On the basis of the observed chemistry of methonium in superacids, as well as from molecular orbital calculations, it is suspected that the C_s form is the preferred configuration [19−27]. This structure is some 2 kcal/mol below the energy level of the C_{4v} form, which in turn is favored over the D_{3h} symmetry by about 8 kcal/mol. Nevertheless, all of these forms will coexist in a dynamic steady state [28]. The existence of all forms is supported by hydrogen-deuterium scrambling observed in superacid solutions of deuterated and undeuterated methane [15−19].

Protolytic attack on saturated hydrocarbons is believed to take place either on the C−H or the C−C bond [29−36]. For

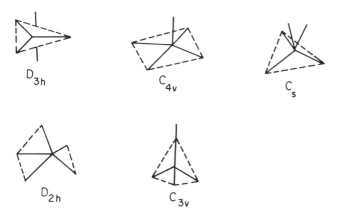

Figure 2.1   Suggested configuration of the methonium ion. (From Ref. 8.)

instance, in the case of methane, the proton attacks the C—H
bond by forcing one of the C—H bonds to share its electron
pair with the electron-deficient proton, as follows:

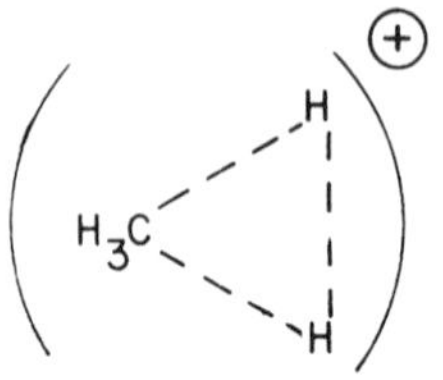

When two or more carbon atoms are present in the molecule, the
proton can interact with either the C—H bonds or the C—C bond
bonds. In the case of ethane, the two possible interactions are
the following:

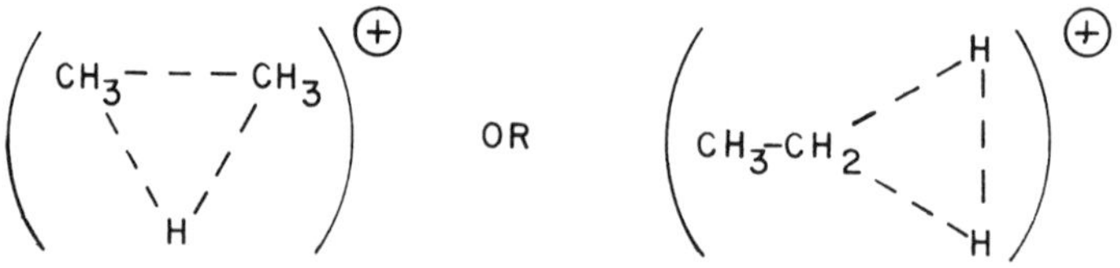

By increasing the chain length of the hydrocarbon, the number
of possible interactions is increased. Since the differences in
free energy between the various species are small, all interactions
are almost equally probable.

In carbenium ions, calculation has shown that the free energy
of ion formation increases with any increase in the number of H
atoms attached to the carbon on which the positive charge is to
appear. One can calculate such energies taking into account the
ionization energies and electron affinities of the hydrogen and
alkyl groups attached to the carbon whose C—H bond is to be
broken. Relative stabilities of some gas-phase carbenium ions
have been calculated in this way many years ago [37] and are
shown in Table 2.1.

The stablilizing effect of the alkyl groups seen in Table 2.1 is
due to a combination of hyperconjugation and the inductive effect.
Based on mass spectrometric data and estimated heats of solvation,
values of the order of 10 kcal/mol have been assigned to the difference in stabilization between tertiary and secondary ions and

Table 2.1 Enthalpies of Gaseous
Ionization $R-H \rightarrow R^{\oplus} + H^{\ominus}$

Type of ion Type of ion	Relative value of $\Delta \Delta H_g$ (kcal/mol)
Methyl	0
Ethyl	-31
i-Propyl	-54
t-Butyl	-70
n-Propyl	-30
n-Butyl	-30
Neophentyl	-43
Allyl	-65

Source: Ref. 37.

25 kcal/mol for that between secondary and primary. The heat evolved in the transformation of a 2-butyl to a *tert*-butyl carbenium ion has been measured to be 14.5 kcal/mol [38]. Ignoring entropy effects, this leads to a room temperature equilibrium of these two species, favoring the tertiary species by a factor of 10^{10}. Clearly, branched ions will dominate in carbenium ions.

Although this pattern is valid for carbenium ions in solution, it must be remembered that in catalytic cracking the carbocations are not isolated but are associated with negatively charged counterions and constrained by the pore structure of the solid catalyst. These will alter the effects observed in solution and lead to an even greater variety of causes and effects. For this reason the extensive studies of carbocation behavior done in solution at moderate temperatures should be taken only as a guide to the behavior of such species in high-temperature catalytic reactions on porous solids.

2.2 MECHANISMS OF CARBOCATION FORMATION

Carbocations are formed by reactions that can be grouped in four main classifications:

1. The addition of a cation to an unsaturated molecule

2. The addition of a proton to a saturated molecule

3. The removal of an electron from an electrically neutral species

4. Heterolytic fission of a molecule

Although catalytic cracking is generally carried on in the presence of solid catalysts, the best examples of carbocation reactions, and the ones that are most clearly defined, are those available in the literature on liquid-phase carbocation processes. Hence we will describe the behavior of carbocations on the basis of evidence available from the liquid phase. Later we attempt to correlate this with what occurs on solid catalysts.

2.2.1   Cation Addition to an Unsaturated Molecule

The addition of a cation to an unsaturated system is best illustrated by considering the addition of a cation to an olefin. The addition will be influenced by the strength of the acid and other factors, such as cation stabilization agents, chemical inertness, and the dielectric constant of the medium. In considering such acid-base reactions and their reactivity, one should consider the nature of both the base (hydrocarbon in this case) and the acid. If the hydrocarbon is not strongly basic, an acid of high strength is necessary to achieve the addition. In the case of protonation of aromatic species, strong acids are required to achieve reactions of this sort:

Using strong acids such as HF—SbF$_5$, it is possible to obtain stable carbocations with compounds of very low basicity, such as benzene or xylene.

The protonation of an olefin proceeds by an attack of the proton on the π electrons of the olefin. This leads in a second step to the formation of a σ bond between one of the carbon

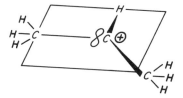

Figure 2.2 Structure of a secondary propyl carbenium ion. (From Ref. 8.)

atoms connected by the double bond and attacking proton. Such bond formation leaves a positive charge on the second carbon atom of the pair, which was initially connected by the double bond.

$$HX + R_1 CH = CHR_2 \rightleftarrows \left[R_1 CH \overset{\overset{\overset{H}{\oplus}}{\vdots}}{=} CHR_2 \right]$$

$$+ X^{\ominus} \rightleftarrows R_1 CH_2 - \overset{\oplus}{C}HR_2 + X^{\ominus}$$

The final carbenium ion produced has an sp^2 hybridization and the configuration around the electron-deficient carbon is planar [39]. The structure of a secondary propyl carbenium ion [40] is illustrated in Figure 2.2. Since the planar configuration is preferred, molecules in which steric hindrance prevents a planar configuration from arising have less stable carbenium ions. Notice also that we are considering a case where the charge is quite clearly localized on one carbon atom.

In the case of the protonation of an aromatic molecule, a similar scheme can be written, but this time the charge is delocalized in the ring structure:

Table 2.2   Heats of Protonation of *m*-Xylene

Molecule	Position of protonation	ΔE_X (atomic units)
m-Xylene	(1), (3)	−0.447
	(2)	−0.475
	(4), (6)	−0.478
	(5)	−0.445

Source: Ref. 51.

Evidence for the protonation of aromatics in superacid media was first obtained in the early 1950s [41–43]. The first spectroscopic evidence of their formation was reported by Gold and Tye [44] and Reid [45]; Doering et al. [46] were first to study benzenium by NMR. It was even found that certain alkylbenzenium ions can be isolated [47–49]. making structural and reactivity studies easier and leaving no doubt as to the existence of such species.

Heats of protonation of methyl-, ethyl-, isopropyl-, and *tert*-butylbenzenes in superacids have been determined calorimetrically [50]. A difference of 3.75 kcal/mol was calculated between *tert*-butyl- and methylbenzene protonation, with ethyl- and propylbenzene falling between the two.

In the protonation of *m*-xylene the following reaction takes place on an acid catalyst:

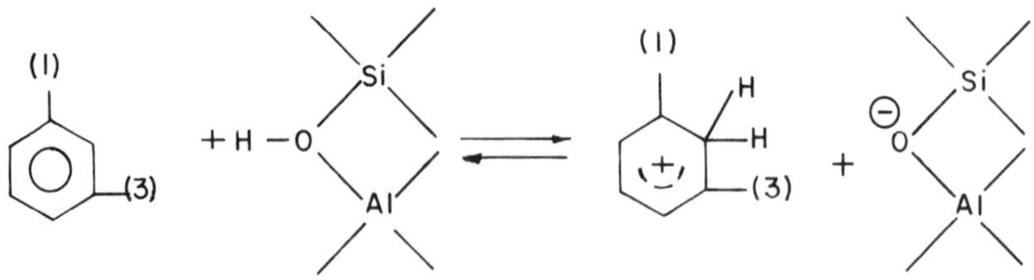

It has been calculated [51] that the energies for the protonation of *m*-xylene in the various positions are as shown in Table 2.2.

From this it can be seen that the preferred positions for protonation are 2, 4, and 6. Thus the center where the proton
attaches itself is dictated by the electron distribution in the
molecule. This, in turn, will lead to selectivity effects in subsequent reactions, since not all carbons in a given molecule are
equally likely to be protonated, nor is the charge equally stable
on the various carbons of an aliphatic molecule.

2.2.2 Proton Addition to a Saturated Molecule

The cracking of alkanes in superacid liquids provides an illustration of proton addition. In this reaction the first step involves
the protonation of the alkane followed by the protolytic cracking
or dehydrogenation of the molecule [32—34]. Several isomeric
structures have been proposed for the carbonium ion resulting
from the protonation of the alkane [52—57]. The interaction of a
proton with a propane molecule was investigated [58] using molecular orbital calculations and the results compared with the experimental work of Hiraoka and Kebarle [55—57].
Various isomeric structures of the protonated propane were
studied for the case of the secondary carbon under attack by a
proton from various starting points. Three structures were
identified as representing minima in the energy surface: a strong
H—H complex with the isopropyl ion, as shown in A; a nonlinear
bridge structure, shown as B; and a linear bridge structure,
shown as C.

In larger hydrocarbons the same three basic structures will be
present, with the additional complication that they will arise on
each of the secondary carbons available.
Complex A represents the attack of a proton on a C—H bond
and has been suspected of taking part in the isomerization of
alkanes in superacid media [32—34]. Such an intermediate will

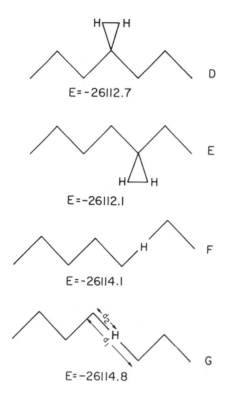

D

E=-26112.7

E

E=-26112.1

F

E=-26114.1

G

E=-26114.8

Figure 2.3   Structures and energies of linear and bridged protonated *n*-heptane.   (From Ref. 35.)

lead to the formation of H_2 and a carbenium ion.   Hydrogen is
known to be evolved during reactions of alkanes in superacids
[32−34].

The other two structures represent an attack on the C—C
bond.   Corma et al. [35] have shown that the linear bridge is the
most stable structure.   Other structures, such as those labeled
D and E, show energy minima similar to those of the nonlinear
bridge structures F and G (see Figure 2.3).

Structure G would be expected to evolve to give

$$C_7H_{17}^{\oplus} \longrightarrow C_3H_8 + C_4H_9^{\oplus}$$

The energy hypersurface corresponding to this cracking process is shown in Figure 2.4. From this we see that the energy barrier for protolytic cracking is in the range 12 to 18 kcal/mol. Moreover, when the bridge structure is formed it is easier for it to evolve via the cracking route than to convert to other structures, since the route of minimum energy leads to cracking [55–57,59].

We conclude that if a catalyst has protons strong enough to interact with alkanes in the way described above, two reactions may take place: the formation of hydrogen by hydride abstraction, or protolytic cracking. The structures leading to cracking are more stable than those leading to the formation of H_2. Thus, in catalytic cracking one may expect that acids on solid materials will generally lead to a paraffin product in the gas phase and a residual carbenium ion on the solid surface. This ion may then initiate further transformations, leading to a kind of chain reaction. Whether hydrogen is also evolved is uncertain. Many authors [35] report the evolution of hydrogen in catalytic cracking on solid catalysts. However, their results are made difficult to interpret by the presence of thermal reactions that are known to produce hydrogen.

2.2.3 Electron Removal from a Neutral Molecule

The generation of carbenium ions from neutral hydrocarbons in the liquid phase was long controversial. The occurrence of the reaction

$$RH + H^{\oplus} \longrightarrow R^{\oplus} + H_2$$

suggested by Bloch et al. [60] remained unproven for many years [61]. However, several authors [33,62–65] have shown that isobutane and other tertiary alkanes react with superacids such as $HF\text{-}SbF_5$ to give stable solutions of the tertiary cations

$$i\text{-}C_4H_{10} + H^{\oplus} \longrightarrow t\text{-}C_4H_9^{\oplus} + H_2$$

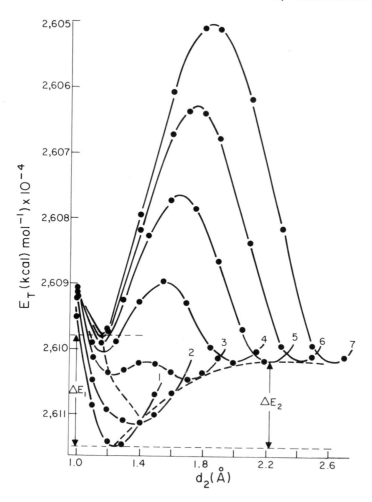

Figure 2.4   Energy corresponding to the cracking of a linear bridged structure of protonated C_7.   Curves in this figure refer to configuration G in Figure 2.3.   Curves labeled 1 to 6 correspond to d_1 lengths of 2.5 to 3.7 Å in increments of 0.2 Å.   Two minima occur in some curves and the loci of the minima are shown by the two dashed lines.   Both minima achieve asymptotic values lying, respectively, ΔE_1 and ΔE_2 above the datum.   The ΔE_1 corresponds to the formation of $C_4^{\oplus}$ and a neutral C_3 fragment, whereas ΔE_2 corresponds to $C_3^{\oplus}$ and a neutral C_4 fragment.

It is now generally accepted that Lewis sites or strong carbenium ions can indeed interact with neutral molecules by abstracting a hydride ion, leaving behind a carbenium ion [7,29,32,34,40,65, 67]. Thus the reactions

$$RH + L \longrightarrow R^{\oplus} + LH^{\ominus}$$

$$RH + R_1^{\oplus} \longrightarrow R^{\oplus} + R_1H$$

are accepted as being possible.

Such reactions have long been suspected of contributing to catalytic cracking by solid catalysts. The catalysts are not only well covered by various carbenium ions during reaction, but even in the initial stages of reaction when no carbenium ions are present, they are known to possess Lewis sites of considerable strength. In view of this a whole new set of reaction possibilities must be considered as part of the catalytic cracking process— most important, as part of the initiation of cracking by Lewis sites.

2.2.4 Heterolytic Fission

The heterolytic fission of a stable molecule involves the breaking up of a molecule into two oppositely charged fragments. If such a breakup takes place with the breaking of a covalent bond, the result is two free radicals. If, however, one of the entities takes two electrons with it, the resultant species have net electrical charges and form an ion pair.

$$RR' \longrightarrow R^{\oplus} + R'^{\oplus}$$

2.3 REACTIONS OF CARBOCATIONS

Regardless of the origins of the carbocation, once it is formed it may undergo any of the following several processes: (1) charge isomerization, (2) chain isomerization, (3) hydride transfer, (4) alkyl transfer, and (5) formation and breaking of carbon-carbon bonds. Although all these reactions are possible, they are not all equally probable for each carbocation. We will therefore examine the foregoing reaction mechanisms with a view to understanding their likelihood of occurrence in a given case.

2.3.1 Charge Isomerization

The isomerization of carbenium ions takes place by hydrogen transfer along the hydrocarbon chain. The difference in energy for the various carbenium ions in a paraffinic chain is small, except for the terminal carbon atoms. For example, in the case of n-heptane [68] the enthalpy of formation for the different carbenium ions is given in Table 2.3.

Sanders and Hagen [69] have studied the activation energy and $\Delta G^{\ddagger}$ for the 1−2 hydrogen shift in the propyl cation.

$$
\begin{array}{ccc}
\quad\;\; H & & \quad\;\; H \\
\quad\;\; | & & \quad\;\; | \\
H_3C - \underset{\oplus}{C} - CH_3 & \rightleftharpoons & H_3C - \underset{|}{C} - \underset{\oplus}{CH_2} \\
& & \quad\;\; H
\end{array}
$$

Table 2.3   Relative Values for Enthalpy of Formation of Carbenium Ions on n-Heptane

Carbenium ion on:	ΔH (kcal/mol)
$C_1 \oplus$	19.1
$C_2 \oplus$	1.5
$C_3 \oplus$	0.6
$C_4 \oplus$	0.0

Source: Ref. 68.

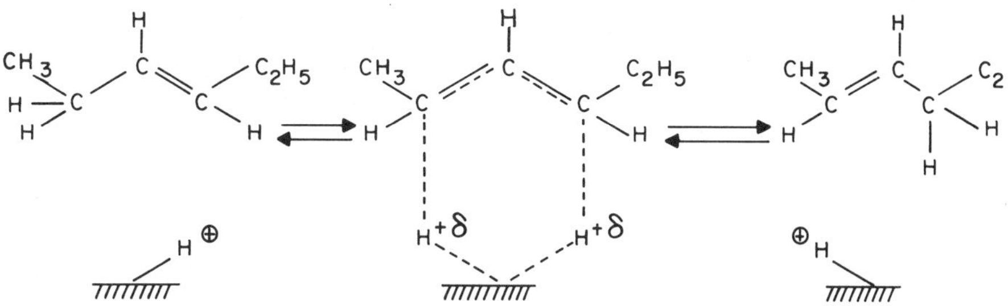

Figure 2.5 Mechanism of double-bond isomerization. (From Ref. 71.)

In principle, the previous reaction could occur by a simple reversible primary—secondary 1—2 hydrogen shift, as evidenced by proton-scrambling studies. However, since the value of ΔG is low compared with the differences in stability between the two species, one has to conclude that the observed reaction does not proceed via a primary cation. In double-bond isomerization, where a similar problem is encountered, several authors [70,71] have postulated a concerted mechanism of the type shown in Figure 2.5.

According to Brouwer [71], catalyst selectivity in this reaction arises from competition between two reactions. One is the complete proton transfer to form a carbenium ion, which leads to cis-trans isomerization and, to a minor extent, to a double-bond shift. The other is the concerted reaction shown in Figure 2.5, leading to a double-bond shift.

Besides the 1—2 hydrogen shift, the more unexpected 1—3 hydrogen shift has been shown to occur [72—75]. For example, the 1—3 hydrogen shift has been observed by proton magnetic resonance (PMR) spectroscopy in the 2,4-dimethylpentyl ion.

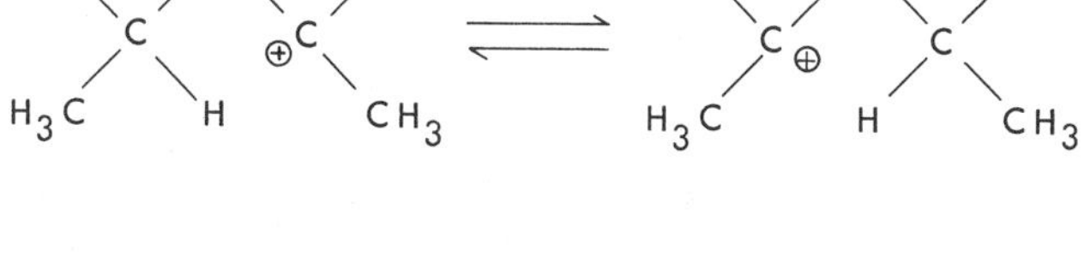

Two possible intermediates can be visualized for this isomerization process, as follows:

No reports of a 1–4 or 1–5 hydrogen shift have been verified to date, but there seems to be no fundamental reason why such shifts should not occur. There is therefore reason to believe that concerted transition states play an important part in this type of reaction.

2.3.2   Skeletal Isomerization

Chain isomerization by a methyl shift generally leads to more highly branched carbenium ions. The rate of such methyl-shift processes is usually about 1000 times smaller than that of the hydrogen shift discussed in the preceding section. The following is a sequence showing a 1–2 hydrogen shift and a 1–2 alkyl shift:

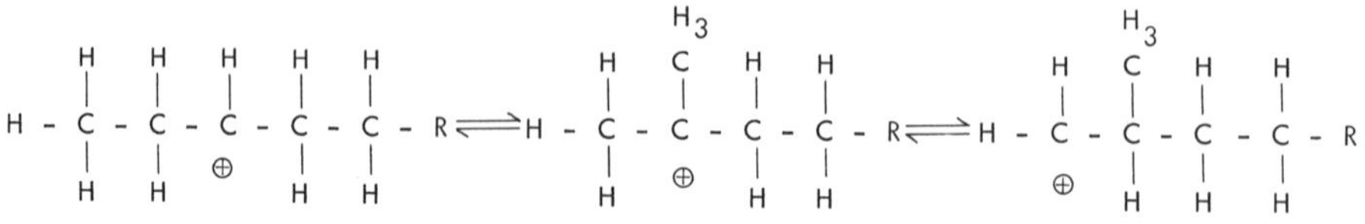

Because a primary carbenium ion is involved in the mechanism shown above, the actual route of the reaction is thought to proceed via intermediates other than those shown above. Brouwer and Oelderik [63] have developed a mechanism that involves a system of protonated cyclopropane rings [36,76] as an intermediate. This not only avoids the intermediacy of a primary carbenium ion, but also explains observations of the isomerization of n-butane-1-^{13}C and of n-pentane catalyzed by superacids [34]. Indeed, n-pentane isomerizes very quickly to isopentane, while n-butane undergoes only a slow isomerization to isobutane under the same conditions. However, n-butane-1-^{13}C is rapidly isomerized to n-butane-2-[13]C.

$$13CH_3 - CH_2 - CH_2 - CH_3 \rightleftharpoons CH_3 - 13CH_2 - CH_2 - CH_3$$

The rate of carbon scrambing in *n*-butane is in fact approximately equal to the rate of isomerization of *n*-pentane to isopentane. All of this can be explained if a protonated cyclopropyl intermediate is postulated:

$$CH_3 - CH \underset{H^{\oplus}}{\overset{13 CH_2}{\triangle}} CH_2$$

Once this intermediate is formed, any carbon-carbon bond in the ring can be broken. The product obtained will depend on which specific bond is broken. Thus the products one would expect are as follows:

$$CH_3 - \overset{\oplus}{\underset{|}{CH}} - CH_2 \quad , \quad CH_3 - \overset{\oplus}{\underset{|}{C}} - CH_2 - 13CH_3 \quad , \quad CH_3 - \overset{\oplus}{\underset{|}{C}} - 13CH_2 - CH_3$$
$$\underset{13CH_3}{} \qquad\qquad \underset{H}{} \qquad\qquad\qquad \underset{H}{}$$

A   B   C

In fact, it is product C that is observed experimentally to be dominant. In the case of the formation of isopentane from *n*-pentane, the mechanism involved would be as follows:

$$CH_3 - \overset{\oplus}{CH} - CH_2 - CH_2 - CH_3 \rightleftharpoons CH_3 - CH \underset{CH_2}{\overset{CH}{\diagup}} \overset{CH_3}{\underset{H^{\oplus}}{}} \rightleftharpoons$$

$$CH_3 - \overset{\overset{CH_3}{|}}{\underset{|}{\overset{\oplus}{C}H}} - CH_3 \rightleftharpoons CH_3 - \overset{\overset{CH_3}{|}}{\underset{\oplus}{C}} - CH_2 - CH_3$$

Another type of isomerization involving a change in the carbon skeleton is that leading to interconversions of isomers having at least one tertiary carbon which is preserved but which changes its position in the molecule [77]:

$$CH_3 - \underset{\underset{\displaystyle CH_3}{|}}{CH} - CH_2 - CH_2 - CH_3 \;\rightleftarrows\; CH_3 - CH_2 - \underset{\underset{\displaystyle CH_3}{|}}{CH} - CH_2 - CH_3$$

These types of isomerization are easier to carry out than those involving changes from secondary to tertiary carbon and can therefore be carried out on less active catalysts. With a catalyst capable of performing both types of isomerizations, the isomerization involving the change in the position of the tertiary carbon is faster than the one that changes the type of carbon. For example, McCaulay [78] has reported that in the presence of HF + 10% BF_3 as catalyst, the following takes place (with the relative rates shown beside the arrows):

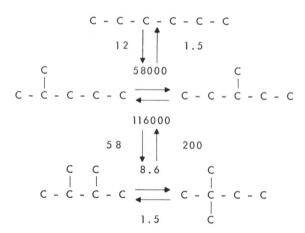

This suggests that the final product distribution in a reaction involving isomerization processes will depend not only on the reaction temperature but also on the activity of the catalysts. It may be possible to produce large changes in selectivity by changing the acid strength of a catalyst. This is especially important when considering solid catalysts whose acid strength distribution can readily be changed by a variety of treatments, as will be discussed later.

Although there is a marked difference in rate between the branching and nonbranching rearrangements, it is believed that both types go through the cyclopropyl carbonium ion stage and that the essential difference between the two types of rearrangements is that the branching rearrangements require higher activities or higher temperatures.

The multiplicity of reactions discussed to this point is well illustrated with one example. The reaction network shown in Figure 2.6 outlines the fate of a hexyl carbenium ion as it isomerizes to the various configurations that lead to the observed products [36].

Another example of such concerted intermediate states can be found in the reactions of cycloalkanes. The isomerization of methylcyclohexane has been studied using the MINDO/3 calculational approach and then comparing the results with those obtained experimentally using an REHY catalyst [79]. The results show that the tertiary carbenium ion X gives the 1,2-dimethylcyclopentyl ions denoted as B and C in Figure 2.7.

The intermediate Y is formed by the 1,2 displacement of the equatorial hydrogen atom bonded to C_β toward the C_α carbon. Calculations show that about 26 kcal/mol is necessary to achieve configuration B, while about 40 kcal/mol is necessary to produce C via a secondary 1,1-dimethylcyclopropyl ion Z.

2.3.3 Hydride Transfer

This important type of reaction can be illustrated as follows:

$$R_1 - H + R_2\overset{+}{} \rightleftarrows R_1\overset{+}{} + R_2 - H$$

Hydride-transfer reactions between alkanes and carbenium ions are important in catalytic cracking of hydrocarbons since they are responsible for a chain process that occurs after the first carbenium ion has formed on a catalyst surface. It has been shown that this reaction is very fast [63,80] and that the rate of abstraction of a secondary hydrogen by a tertiary carbenium ion is roughly the same for all normal alkanes [36]. The tertiary-tertiary hydride transfer is even faster than the secondary-tertiary, which in turn

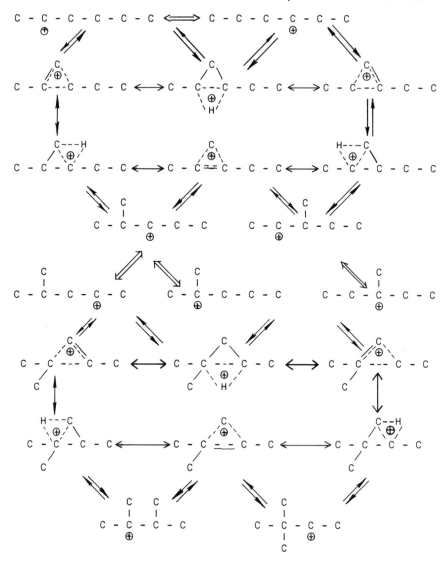

Figure 2.6   1-2 Hydrogen shifts (double arrows) and alkylcyclo-propyl ion interconversions (single arrows).   (From Ref. 36.)

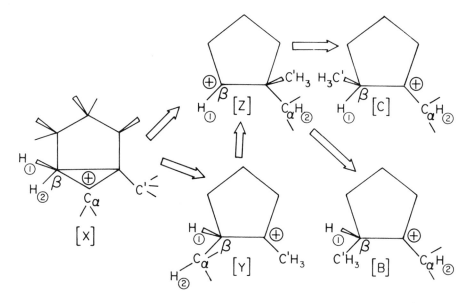

Figure 2.7 Isomerization of methylcyclohexane. (From Ref. 79.)

is faster than the primary-secondary transfer [81]. This shows that hydride abstraction is accelerated by neighboring groups, which encourage the stabilization of the resultant ion [36]. An important question in this case is the following: If a carbenium ion is able to abstract a hydride from a paraffin, is it able to react with a molecule of hydrogen? Such a reaction would deactivate the carbenium ion by forming a paraffin and would leave a proton on the surface. It has been shown by molecular orbital calculations that this reaction will occur in the gas phase [27], in the liquid phase in the presence of HF-SbF$_5$ [83,84], and may be present in heterogeneous acid catalysis [85]. This means that in catalytic cracking systems the presence of molecular hydrogen in such reactions will have an effect on the paraffin-to-olefin ratio and, more important, on coke formation.

It seems that hydride-transfer reactions are easy and that their preferred products are the same as those expected from internal hydrogen migrations. The main difference is that by hydride transfer a charge is passed from one molecule to another, thus propagating a chain reaction. Such chains may be important in coke formation and catalyst decay, as will be seen later.

2.3.4   Alkyl Transfer

Alkyl-transfer reactions can be represented by

$$R_1^{\oplus} + R_2 - R_3 \longrightarrow R_1 - R_2 + R_3^{\oplus}$$

Such displacement reactions have not been proven directly. However, an attempt has been made to establish their existence, as reported by Hogeveen and Bickel [86] for the following reaction:

$$\underset{\substack{CH_3 \\ | \\ CH_3 - C^{\oplus} \\ | \\ CH_3}}{} + \underset{\substack{CH_3 \\ | \\ H_3C - C - CH_2D \\ | \\ CH_3}}{} \rightleftharpoons \underset{\substack{CH_3 \\ | \\ H_3C - C - CH_3 \\ | \\ CH_3}}{} + \underset{\substack{CH_3 \\ | \\ {}^{\oplus}C - CH_2D \\ | \\ CH_3}}{}$$

Reactions of this kind, if they occur, do not seem to have any important consequences in catalytic cracking. A similar process, called disproportionation, is more important and will be discussed below.

2.3.5   Carbon-Carbon Bond Formation and Breaking

Carbocations are important intermediates in reactions involving the formation and breaking of carbon-carbon bonds. The cracking process, which is responsible for the majority of motor fuels produced from crude petroleum, is largely dependent on the bond-breaking processes that occur on acid catalysts. Two main groups of bond-forming reactions, olefin polymerization and the alkylation of paraffins and aromatics, are also considered here. An important variant of these two reactions involves both the formation of carbon-carbon bonds and their breaking: this is the disproportionation reaction.

There is much evidence that the presence of carbocations is essential in certain liquid-phase polymerization reactions [36]. In general, such polymerization reactions involve several consecutive steps which can be summarized as follows:

$$\text{Initiation} \quad \overset{|}{\underset{|}{C}} = \overset{|}{\underset{|}{C}} \quad + \quad HX \rightleftharpoons \overset{|}{\underset{|}{C}} - \overset{|}{\underset{\oplus}{C}} \quad\longrightarrow\; + \; X^{\ominus}$$

$$\text{Propagation} \quad R^{\oplus} \; + \quad \overset{|}{\underset{|}{C}} = \overset{|}{\underset{|}{C}} \longrightarrow R - \overset{|}{\underset{|}{C}} - \overset{|}{\underset{|}{C}} \; \oplus$$

$$\text{Termination} \quad R^{\oplus} + X^{\ominus} \longrightarrow RX$$

Clearly, processes such as cracking, hydride abstraction, skeletal rearrangment, and other reactions already mentioned can take place during this kind of polymerization, increasing the complexity of the resultant polymer products. High-molecular-weight polymers are normally formed by ionic polymerizations at low temperatures where the propagation is much more rapid than cracking or skeletal rearrangement. From our previous discussion we can surmise that during the formation of the polymer, secondary carbenium ions predominate over the less stable primary ions, with the result that highly branched polymers are formed by this reaction [40].

Carbon-carbon bond formation also occurs in alkylation. For example, the alkylation of an aromatic occurs by the attack of a carbenium ion on the π electrons of the benzene ring.

The product ion can be thought of as a σ complex which readily gives up a proton from the tetrahedral carbon, leaving the alkylated aromatic. Kinetic studies of aromatic alkylation indicate the rate-determining step in this process is the formation of the σ complex [87,88]. On a solid catalyst the alkylation reaction in all probability takes place via a Rideal mechanism [89]. This mechanism presupposes the adsorption of the alkylating agent on an active site on the catalyst surface. In the case of a solid-acid catalyst, the active site is a Brønsted acid site and the adsorbed product is the carbenium ion.

$$CH_2 = CH_2 + H^{\oplus\delta} - O^{\ominus\delta} - \text{Surface} \underset{k_2}{\overset{k_1}{\rightleftharpoons}} CH_3 - CH_2^{\oplus} - O^{\ominus} - \text{Surface}$$

If the surface carbenium ion equilibrium is readily established, the rate of the overall reaction is governed by the rate of the attack by the aromatic ring on the carbenium ion:

not by the rate of carbenium ion formation. This is followed by a desorption of the product, as follows:

with the resultant formation of the substituted aromatic and the regeneration of the original Brønsted acid site.

The alkylation of isoparaffins by olefins is also carried out in the presence of an acid catalyst. The reaction mechanism involves the formation of a carbenium ion by the addition of a proton to the double bond of the olefin, followed either by a polymerization of the olefins present or by a hydride ion extraction from the paraffins present in the reaction medium. The resultant product depends on the relative specific rates of polymerization and of hydride ion transfer, as well as on the concentrations of the isoparaffins and olefins present in the reacting mixture. To increase the rate of hydride transfer, paraffins with tertiary carbon atoms such as isobutane are used in the alkylation process. The overall process is as follows:

Formation of the carbenium ion:

$$H_2C = CH - CH_3 + HX \rightleftharpoons CH_3 - \overset{\oplus}{C}H - CH_3 + X^{\ominus}$$

Hydrogen-transfer reaction:

$$CH_3 - \underset{\oplus}{C}H - CH_3 + CH_3 - \underset{\underset{H}{|}}{\overset{\overset{CH_3}{|}}{C}} - CH_3 \rightleftharpoons CH_3 - CH_2 - CH_3 + CH_3 - \underset{\oplus}{\overset{\overset{CH_3}{|}}{C}} - CH_3$$

Alkylation reaction:

$$CH_3 - \underset{\oplus}{\overset{\overset{CH_3}{|}}{C}} - CH_3 + CH_2 = CH - CH_3 \rightleftharpoons CH_3 - \overset{\overset{CH_3}{|}}{\underset{\underset{\underset{\underset{CH_3}{|}}{\overset{\oplus}{C}H}}{\overset{|}{CH_2}}}{C}} - CH_3$$

Termination reaction:

$$CH_3 - \overset{\overset{CH_3}{|}}{\underset{\underset{\underset{CH_3}{|}}{\overset{\oplus}{C}H}}{\overset{|}{CH_2}}}{C} - CH_3 + CH_3 - \underset{\underset{H}{|}}{\overset{\overset{CH_3}{|}}{C}}H - CH_3 \rightleftharpoons$$

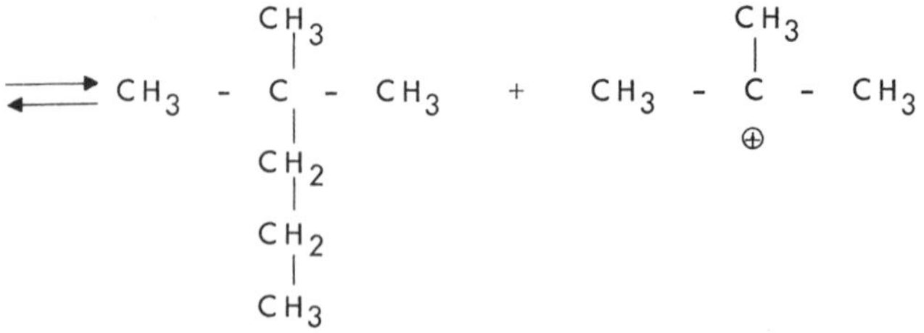

$$\xrightleftharpoons{\quad} \begin{array}{c} CH_3 \\ | \\ CH_3 - C - CH_3 \\ | \\ CH_2 \\ | \\ CH_2 \\ | \\ CH_3 \end{array} \quad + \quad \begin{array}{c} CH_3 \\ | \\ CH_3 - \overset{\oplus}{C} - CH_3 \end{array}$$

Clearly, a high ratio of paraffin to olefin in the feed will enhance the rate of alkylation by the foregoing scheme and reduce the rate of polymerization. In commercial alkylation processes, branched heptanes are obtained and the resultant gasoline is prized for its high octane number.

Side reactions are often a source of problems. For instance, in this reaction the alkylcarbenium ions can also react with olefins to give polyalkylcyclopentyl cations and alkanes [90].

$$i\text{-}C_4H_8 \;+\; H^{\oplus} \;\xrightleftharpoons{\quad}\; t\text{-}C_4H_9^{\oplus}$$

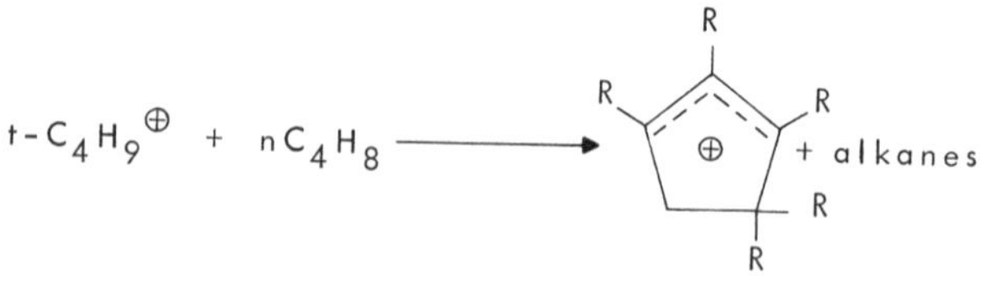

$$t\text{-}C_4H_9^{\oplus} \;+\; nC_4H_8 \;\longrightarrow\; \text{(ring)} \;+\; alkanes$$

Such reactions deactivate the catalyst. The reaction is not simple but involves many steps and illustrates the potential complexities once an active species such as a carbenium ion is formed.

The cracking reaction is the reverse of the types shown above and takes place by the breaking of carbon-carbon bonds which are β to the position of the carbon containing the charge. This reaction is discussed in more detail in subsequent sections; here it will suffice to give the general rules that govern the cracking reaction.

1. The bond that is broken is located in the β position to the carbon atom with the positive charge, that is,

$$R' - \overset{\overset{\displaystyle H}{|}}{\underset{\underset{\displaystyle H}{|}}{C}} - \overset{\overset{\displaystyle H}{|}}{\underset{\underset{\displaystyle H}{|}}{C}} - \overset{\oplus}{\underset{\underset{\displaystyle H}{|}}{C}} - \overset{\overset{\displaystyle H}{|}}{\underset{\underset{\displaystyle H}{|}}{C}} - \overset{\overset{\displaystyle H}{|}}{\underset{\underset{\displaystyle H}{|}}{C}} - R$$

$$R' - \overset{\overset{\displaystyle H}{|}}{\underset{\underset{\displaystyle H}{|}}{C}}^{\oplus} + \overset{\overset{\displaystyle H}{|}}{\underset{\underset{\displaystyle H}{|}}{C}} = \overset{\overset{\displaystyle H}{|}}{C} - \overset{\overset{\displaystyle H}{|}}{\underset{\underset{\displaystyle H}{|}}{C}} - \overset{\overset{\displaystyle H}{|}}{\underset{\underset{\displaystyle H}{|}}{C}} - R$$

or

$$R' - \overset{\overset{\displaystyle H}{|}}{\underset{\underset{\displaystyle H}{|}}{C}} - \overset{\overset{\displaystyle H}{|}}{\underset{\underset{\displaystyle H}{|}}{C}} - \overset{\overset{\displaystyle H}{|}}{C} = \overset{\overset{\displaystyle H}{|}}{C} + \overset{\overset{\displaystyle H}{|}}{\underset{\underset{\displaystyle H}{|}}{C}}^{\oplus} - R$$

2. The product olefins are 1-olefins and are released to the gas phase by the cracking event, leaving on the surface a smaller carbenium ion.

3. Cracking takes place in such a way that the dominant carbenium ion produced is the more stable of the two possible, that is,

$$CH_3 - CH_2 - \overset{\oplus}{CH} - CH_2 - CH_2 - R$$

No → $\overset{\oplus}{CH_3} + CH_2 = CH - CH_2 - CH_2 - R$

Yes → $CH_3 - CH_2 - CH = CH_2 + \overset{\oplus}{CH_2} - R$

4. The product carbenium ion may desorb or isomerize to a more stable configuration or crack again.

The illustrative examples given above will in practice not be the favored reactions since they require the formation of primary carbenium ions. A route that involves initial isomerization to produce branched species which then crack without producing a primary carbenium ion is much more likely. For example, in the case of octanes, Brouwer [82] presents the following scheme:

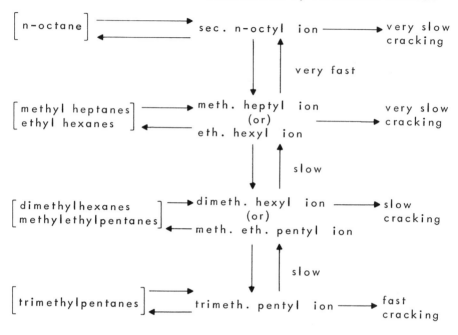

The last reaction to be considered under this heading is the disproportionation reaction [91-98]. An example of a disproportionation is the overall process

$$2C_5H_{12} \rightleftarrows C_4H_{10} + C_6H_{14}$$

This reaction proceeds through a complex sequence of steps involving the alkylation of one C_5 species by a $C_5^{\oplus}$ carbenium ion which leads to a carbenium ion of twice the original length. This is followed by the scission of an internal carbon-carbon bond, leading to the disproportionation shown [98]. Reactions such as this have been reported in the cracking of *n*-heptane on a chromium zeolite CrHNay [97] and in the cracking of olefins [99]. Disproportionation reactions have also been reported to be important in the cracking of alkyl aromatics, as for example in the cracking of cumene [100,101]:

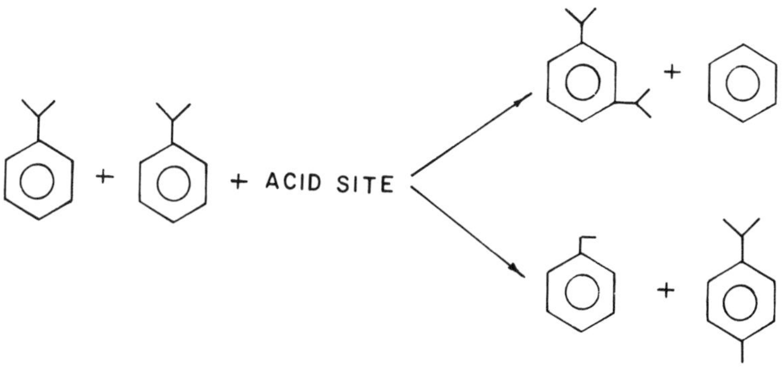

The various reactions involved in cumene cracking are a good illustration of many of the processed discussed above and are presented in some detail in Chapter 5.

2.4 CONCLUSION

When a hydrocarbon reacts on the surface of a solid catalyst it is seldom a straightforward process. A series of reactions occurs. All involve a carbocation intermediate and take place at rates which are governed both by the nature of the carbocation formed and by the nature and strength of the acid site involved in the catalysis. The schematic representation in Figure 2.8 indicates the relative acid strengths required for the various reactions mentioned above.

The relative rates of the various reactions, taken together with the relative ease of formation of the various carbocations from the parent molecule, lead to a bewildering array of possible reaction paths. The matter is complicated by the presence of a variety of active sites on heterogeneous catalysts. These sites not only differ in acid strength but may also, as will be seen later, differ in nature. Furthermore, the picture is complicated by the presence of a solid surface which can induce steric phenomena, and by the possibilities of diffusional resistance which can seriously distort the kinetic picture being observed.

Evidence has been reported which suggests that any given reaction occurs on a specific narrow range of active site acidities

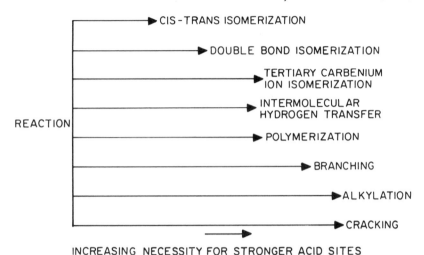

INCREASING NECESSITY FOR STRONGER ACID SITES

Figure 2.8   Relative strength of acid sites required in carbenium ion reactions.

[102].   Thus by adjusting the distribution of site acidities one should be able to control the selectivity of cracking or other carbocation reactions occurring on heterogeneous catalysts.

Since the strength, nature, and distribution of active sites as well as the diffusional characterisitics of catalysts are a function of the method of catalyst preparation, our next topic concerns the nature, preparation, and evaluation of cracking catalysts.

REFERENCES

1.   D. B. Bethell and V. Gold, *Carbonium Ions; An Introduction*, New York, Academic Press, 1967

2.   J. Schmidlin, *Das Triphenylmethyl*, Stuttgart, Enke, 1914

3.   H. Meerwein and K. van Emster, *Ber.*, 55: 2500 (1922)

4.   F. C. Whitmore, *J. Am. Chem. Soc.*, 54: 3274 (1932)

5.   G. A. Olah, *Chem. Eng. News*, 45: 77 (Mar. 27, 1967); *Science, 168*, 1298 (1970)

6.   G. A. Olah and K. Laali, *J. Org. Chem.*, 49: 4591 (1984)

7. G. A. Olah, *Angew. Chem. Int. Ed. Engl.*, *12*: 173 (1973)

8. G. A. Olah, *J. Am. Chem. Soc.*, *94*: 808 (1972)

9. G. A. Olah and J. A. Olah, *Carbonium Ions* (G. A. Olah and P. R. Schleyer, Eds.), Vol. II, p. 715, New York, Wiley, 1970

10. G. D. Sargent, *Q. Rev. Chem. Soc.*, *20*: 301 (1966)

11. S. Winstein, *Q. Rev. Chem. Soc.*, *23*: 14 (1969)

12. H. C. Brown, *Chem. Eng. News*, *45*: 87 (Feb. 13, 1967)

13. V. L. Tal'roze and A. K. Lyubimova, *Dokl. Akad. Nauk SSSR*, *86*: 909 (1952)

14. F. H. Field and M. S. B. Munson, *J. Am. Chem. Soc.*, *87*: 3289 (1965)

15. G. A. Olah and J. Lukas, *J. Am. Chem. Soc.*, *89*: 2227 (1967); *89*: 4739 (1967); *90*: 933 (1968)

16. H. Hogeveen and A. F. Bickel, *Chem. Commun.*, *3*: 635 (1967)

17. H. Hogeveen and C. J. Gaasbeek, *Recl. Trav. Chim. Pays-Bas*, *87*: 319 (1968)

18. H. Hogeveen, C. J. Gaasbeek, and A. F. Bickel, *Recl. Trav. Chim. Pays-Bas*, *88*: 703 (1969)

19 G. A. Olah, G. Klopman, and R. H. Schlosberg, *J. Am. Chem. Soc.*, *91*: 3261 (1969)

20. A. Gamba, G. Morosi, and M. Simonetta, *Chem. Phys. Lett.*, *3*: 20 (1969)

21. W. Th. A. M. van der Lugt and P. Ros, *Chem. Phys. Lett.*, *4*: 389 (1969)

22. J. J. C. Mulder and J. S. Wright, *Chem. Phys. Lett.*, *5*: 445 (1970)

23. H. Kollmar and H. O. Smith, *Chem. Phys. Lett.*, *5*: 7 (1970)

24. V. Dyczmons, V. Staemmler, and W. Kutzelnigg, *Chem. Phys. Lett.*, *5*: 361 (1970)

25. W. A. Lathan, W. J. Hehre, and J. A. Pople, *Tetrahedron Lett.*, 2699 (1970)

26. W. A. Lathan, W. J. Hehre, and J. A. Pople, *J. Am. Chem. Soc.*, *93*: 808 (1971)

27. A. Corma, J. Sanchez, and F. Tomas, *J. Mol. Catal.*, *19*: 9 (1983)

28. E. L. Muetterties, *J. Am. Chem. Soc.*, *91*: 1636 (1969)

29. G. A. Olah, Y. Halpern, J. Shen, and Y. K. Mo, *J. Am. Chem. Soc.*, *93*: 1251 (1971)

30. G. A. Olah and J. A. Olah, *J. Am. Chem. Soc.*, *93*: 1256 (1971)

31. G. A. Olah and H. C. Lin, *J. Am. Chem. Soc.*, *93*: 1259 (1971)

32. G. A. Olah, J. R. DeMember, and J. Shen, *J. Am. Chem. Soc.*, *95*: 4952 (1973)

33. G. A. Olah, Y. Halpern, J. Shen, and Y. K. Mo, *J. Am. Chem. Soc.*, *95*: 4960 (1973)

34. G. A. Olah, Y. K. Mo, and J. A. Olah, *J. Am. Chem. Soc.*, *95*: 4939 (1973)

35. A. Corma, J. H. Planelles, J. Sanchez-Martin, and F. Tomas, *J. Catal.*, *92*: 284 (1985)

36. D. M. Brouwer and H. Hogeveen, *Prog. Phys. Org. Chem.*, *9*: 179 (1972)

37. H. O. Pritchard, *Chem. Rev.*, *52*: 529 (1953)

38. E. W. Bittner, E. M. Arnett, and M. Saunders, *J. Am. Chem. Soc.*, *98*: 3734 (1976)

39. D. H. Olson, *J. Phys. Chem.*, *72*: 1400 (1968)

40. B. C. Gates, J. R. Katzer, and G. C. A. Schmit, *Chemistry of Catalytic Processes*, New York, McGraw-Hill, 1979

41. D. A. McCaulay and A. P. Lien, *J. Am. Chem. Soc.*, *73*: 2013 (1951)

42. H. C. Brown and H. W. Pearsall, *J. Am. Chem. Soc.*, *73*: 4681 (1951); *74*: 191 (1952)

43. H. C. Brown and W. J. Wallace, *J. Am. Chem. Soc.*, *75*: 6268 (1953)

44. V. Gold and F. L. Tye, *J. Chem. Soc.*, 2172 (1952)

45. C. Reid, *J. Am. Chem. Soc.*, *76*: 3264 (1954)

46. W. von E. Doering, M. Saunders, H. G. Boyton, J. W. Earhart, E. F. Wadley, W. R. Edwards, and G. Laber, *Tetrahedron*, *4*: 178 (1958)

47. G. A. Olah, S. J. Kuhn, and A. Pavlath, *Nature, 178*: 693 (1956)

48. G. A. Olah and S. J. Kuhn, *J. Am. Chem. Soc., 80*: 6535 (1958)

49. G. A. Olah, R. H. Schlosberg, R. D. Porter, Y. K. Mo, D. P. Kelly, and G. D. Mateescu, *J. Am. Chem. Soc., 94*: 2034 (1972)

50. E. M. Arnett and J. W. Larsen, *J. Am. Chem. Soc., 91*: 1438 (1969)

51. A. Corma, A. Cortes, I. Nebot, and F. Tomas, *J. Catal., 57*: 444 (1979)

52. R. A. Poirier, E. Constantin, J. Ch. Abbe, M. R. Pearson, and I. G. Csizmadia, *J. Mol. Struct., 88*: 343 (1982)

53. K. B. Lipkowitz, R. M. Larter, and D. B. Boyd, *J. Am. Chem. Soc., 102*: 85 (1980)

54. K. Raghavachari, R. A. Whiteside, J. A. Pople, and P. R. Schleyer, *J. Am. Chem. Soc., 103*: 5649 (1981)

55. K. Hiraoka and P. Kebarle, *J. Chem. Phys., 63*: 394 (1975)

56. K. Hiraoka and P. Kebarle, *Can. J. Chem., 53*: 970 (1975)

57. K. Hiraoka and P. Kebarle, *J. Am. Chem. Soc., 98*: 6119 (1976)

58. J. H. Planelles, Ph.D. thesis, Universidad de Valencia, 1984

59. J. H. Planelles, J. Sanchez-Martin, and F. Tomas, *J. Mol. Struct. Theo. Chem., 108*: 65 (1984)

60. H. S. Bloch, H. Pines, and L. Schmerling, *J. Am. Chem. Soc., 68*: 153 (1946)

61. C. P. Brewer and B. S. Greensfelder, *J. Am. Chem. Soc., 73*: 2257 (1951)

62. H. Hogeveen and A. F. Bickel, *Recl. Trav. Chim. Pays-Bas, 86*: 1313 (1967)

63. D. M. Brouwer and J. M. Oelderik, *Recl. Trav. Chim. Pays-Bas, 87*: 721 (1968)

64. H. Hogeveen, C. J. Gaasbeek, and A. F. Bickel, *Recl. Trav. Chim. Pays-Bas, 88*: 703 (1969)

65. H. H. Voge, *Catalysis*, Vol. VI, (P. H. Emmett, Ed.),
    p. 407, New York, Reinhold, 1958

66. H. Pines and N. E. Hoffman, *Friedel-Crafts and Related
    Reactions*, Vol. II(2), p. 1121, New York, Interscience, 1964

67. H. Hattori, O. Takahashi, M. Takagi, and K. Tanabe,
    J. Catal., *68*: 132 (1981)

68. A. Corma, A. Lopez Agudo, I. Nebot, and F. Tomas,
    J. Catal., *77*: 159 (1982)

69. M. Saunders and E. L. Hagen, *J. Am. Chem. Soc.*, *90*: 6881
    (1968)

70. J. Turkevich and R. K. Smith, *J. Chem. Phys.*, *16*: 466
    (1948)

71. D. M. Brouwer, *J. Catal.*, *1*: 22 (1962)

72. O. A. Reutov and T. N. Shatkina, *Dokl. Akad. Nauk SSSR*,
    133: 606 (1960)

73. P. S. Skell and R. J. Maxwell, *J. Am. Chem. Soc.*, *84*:
    3963 (1962)

74. A. Reutov and T. N. Shatkina, *Tetrahedron Lett.*, *18*: 237
    (1962)

75. D. M. Brouwer and J. A. Van Doorn, *Recl. Trav. Chim.
    Pays-Bas*, *88*: 573 (1969)

76. M. Saunders, P. Vogel, E. L. Hagen, and J. Rosenfeld,
    Acc. Chem. Res., *6*: 53 (1973)

77. C. D. Nenitzescu, *Carbonium Ions*, (G. A. Olah and P. R.
    Schleyer, Eds.), Vol. II, p. 463, New York, Wiley, 1970

78. D. A. McCaulay, *J. Am. Chem. Soc.*, *81*: 6437 (1959)

79. A. Corma, I. Nebot, P. Viruela, and F. Tomas (to be pub-
    lished)

80. P. D. Barlett, F. E. Condon, and A. Schneider, *J. Am.
    Chem. Soc.*, *66*: 1531 (1944)

81. G. M. Kramer, *J. Am. Chem. Soc.*, *91*: 4819 (1969)

82. D. M. Brouwer, *Chemistry and Chemical Engineering of Cat-
    alytic Processes* (R. Prins and G. C. A. Schuit, Eds.), p. 154,
    Rockville, Md., Sijthoff and Nordhoff, 1980

83. H. Hogeveen and A. F. Bickel, *Recl. Trav. Chim. Pays-Bas*,
    86: 1313 (1967)

84. A. F. Bickel, C. J. Gaasbeek, H. Hogeveen, J. M. Oelderik, and J. C. Platteeuw, *Chem. Commun.*, 634 (1967)

85. Kh. Minachev, V. Garanin, T. Isakova, V. Karlamov, and V. Bogomolov, *Adv. Chem. Ser.* (Mol. Sieve Zeolites—I), *102*: 441 (1971)

86. H. Hogeveen and A. F. Bickel, *Recl. Trav. Chim. Pays-Bas*, *88*: 371 (1969)

87. G. A. Olah, S. H. Flood, and M. E. Moffatt, *J. Am. Chem. Soc.*, *86*: 1060 (1964); *86*: 1065 (1964)

88. G. A. Olah, S. H. Flood, S. J. Kuhn, M. E. Moffat, and N. A. Overchuk, *J. Am. Chem. Soc.*, *86*: 1046 (1964)

89. P. B. Venuto, *Adv. Chem. Ser.* (Mol. Sieve Zeolites—II), *102*: 260 (1971)

90. N. C. Deno, D. B. Boyd, J. D. Hodge, C. U. Pittman, Jr., and J. O. Turner, *J. Am. Chem. soc.*, *86*: 1745 (1964)

91. F. E. Condon, *Catalysis* (P. H. Emmett, Ed.), Vol. II, p. 43, New York, Reinhold, 1958

92. E. O. Box, Jr., *U. S. Patent 3,446,868*, 1969

93. N. Y. Chen and S. J. Lucki, *U. S. Patent 3,812,199*, 1974

94. N. Y. Chen and E. Bowes, *U. S. Patent 3,914,331*, 1975

95. K. Tanabe and H. Hattori, *Chem. Lett.*, *6*: 625 (1976)

96. Y. Ono, T. Tanabe, and N. Kitajima, *J. Catal.*, *56*: 47 (1979)

97. A. Lopez Agudo, A. Asensio, and A. Corma, *J. Catal.*, *69*: 274 (1981)

98. G. A. Fuentes and B. C. Gates, *J. Catal.*, *76* 440 (1982)

99. J. Abbot and B. W. Wojciechowski, *Can. J. Chem. Eng.*, *63*: 278, 451, 462 (1985)

100. H. Pines, *The Chemistry of Catalytic Hydrocarbon Conversions*, New York, Academic Press, 1981

101. A. Corma and B. W. Wojciechowski, *Catal. Rev. Sci. Eng.*, *24*: 1 (1982)

102. K. Tanabe, *Solid Acids and Bases*, New York, Academic Press, 1970

3

Cracking Catalysts

3.1   INTRODUCTION

It is now generally accepted that to crack hydrocarbons, one must
use an acid catalyst, that is, a catalyst capable of producing carbocations on its surface.   Acids in solution could be used if their
acid strength were high enough.   However, technical problems such
as corrosion, separation of phases, and recovery of the catalyst
make liquid or homogeneous catalytic cracking impractical.   Homogeneous catalysts such as metal halides and in particular aluminum
trichloride were in fact tested on a commercial basis and abandoned
because of operating difficulties and severe losses of the aluminum
trichloride in the tarry residues produced [1].
     An alternative to homogeneous cracking catalysts is the hetero-
geneous cracking catalyst.   The first such materials used as catalysts were natural clays.   Unfortunately, despite their activity,
such catalysts are very rapidly deactivated and it was not until
Houdry developed a continuous regeneration process that a practical technical solution to the cracking of petroleum was achieved.
     Once catalyst regeneration by the burning off of the carbona-
ceous residues formed during the cracking reaction was achieved,
significant effort was devoted to the improvement of the available
natural catalysts [2].   The first improvement was achieved by the
acid treatment of natural clays.   However, it was soon found that
artificial clays, such as amorphous synthetic combinations of silica
and alumina, silica magnesia, silica zirconia, and so on, also had
interesting catalytic properties.   Although more costly than the
natural materials, the synthetic silicates had a higher activity and
gave a better product distribution.   In the end, of all the available silicates, silica-alumina proved to be the most interesting composition.   Silica by itself shows no activity or acidity;   however,
as soon as small amounts of alumina are introduced into the mixture, acidity begins to rise and the activity of the material as a
cracking catalyst is improved [3].   We begin our discussion of
cracking catalysts by considering the nature of the amorphous
silica aluminas.

3.2   AMORPHOUS SILICA ALUMINAS

Several methods have been reported in the literature for the
preparation of amorphous silica-alumina cracking catalysts [3,4].
One preparation of such a catalyst involves the combination of a

silica hydrogel with an aluminum sulfate solution, followed by the hydrolysis and precipitation of the aluminum salt by the addition of aqueous ammonia. The resultant silica-alumina hydrogel is washed, dried, and calcined in the desired shape for use as a catalyst. Another type of preparation involves the interaction of sodium silicate and sodium aluminate followed by the exchange of sodium by ammonium ions. This ammonium in turn is removed by high-temperature calcination after drying of the initial material. Other less common methods involve the preparation of a catalyst by the intimate mixing of silica hydrogel and aluminum hydroxide, by the thermal treatment of an aluminum salt on the surface of a silica zerogel, or by the hydrolysis of mixtures of ethyl silicate and aluminum isopropylate.

The many variables that can be introduced during the preparations of a cracking catalyst have a great bearing on the characteristics of the catalyst produced. For example, an excess of reactive silica gel and a pH as low as 3 favor the desired Al—O—Si bond formation rather than the formation of Al—O—Al bonds [5]. At the same time, low pH values favor the formation of high-density low-surface-area material [6]. Other important factors, such as the aging time of the gel and concentrations in the initial mixture, have an influence on the characteristics of the final material [7].

Cations other than Na^+ also stabilize tetrahedrally coordinated aluminum, which appears to be the active form. If bulky organic cations, such as tetraalkyl ammonium cations, are used, the Al(IV) ions are forced apart by steric interference and/or electrostatic effects. If the separation can be maintained during gelling, washing, drying, and activation, silica aluminas are produced in which the active-site separation is dependent on the size of the cation used in the synthesis [8].

One of the properties that has a great influence on catalyst behavior in practice is the pore-size distribution. Efforts have therefore been made to develop silica-alumina catalysts with controlled pore-size distributions. The most important methods used for this can be summarized as follows:

1. The control of the aging step [9] in the standard technique, as described by Plank and Drake [7]

2. Co-gelation of silica and alumina hydrosols in the presence of potassium ions [10]

3. Hydrolysis of bulky organosilanes in the presence of aluminum salts [1]

4. Utilization of tetraalkyl ammonium cations such as tetrabutyl or tetramethyl as the counterions to the aluminate ions in the gels [8]

Other methods are used to control the abrasion resistance of the material, thermal stability, and other properties, resulting in a catalyst that represents the optimum of many characteristics necessary for successful application.

3.2.1 Surface Characteristics of Amorphous Silica Aluminas

Amorphous silica-alumina catalysts are distinguished by the fact that they have no long-range structure and thus appear amorphous to x-ray crystallography. Their structure consists of a random three-dimensional network of interconnected silica and alumina tetrahedra. It is to be expected that in such a random arrangement, a given aluminum atom may have other aluminum atoms as well as silicon atoms as immediate neighbors via oxygen bridges. In those cases when an aluminum atom finds itself forced by neighboring atoms to form a tetrahedral structure via oxygen bridges, it develops a positive charge. These charges are the source of the acid sites responsible for catalytic activity in carbocation reactions.

There is the possibility that surface composition is not the same as that of the bulk phase. Planck and Drake [7] reported that in an 11-wt % aluminum commercial catalyst, most of the aluminum is situated on the surface of the catalyst, whereas others [12–16] say that there is no relative enrichment of SiO_2 or Al_2O_3 near the surface of the silica alumina. Regardless, the surface structure in such material is thought to be something like the following:

Peri [17] attempted to systematize structures that can exist in amorphous silica aluminas, proposing the structures shown in

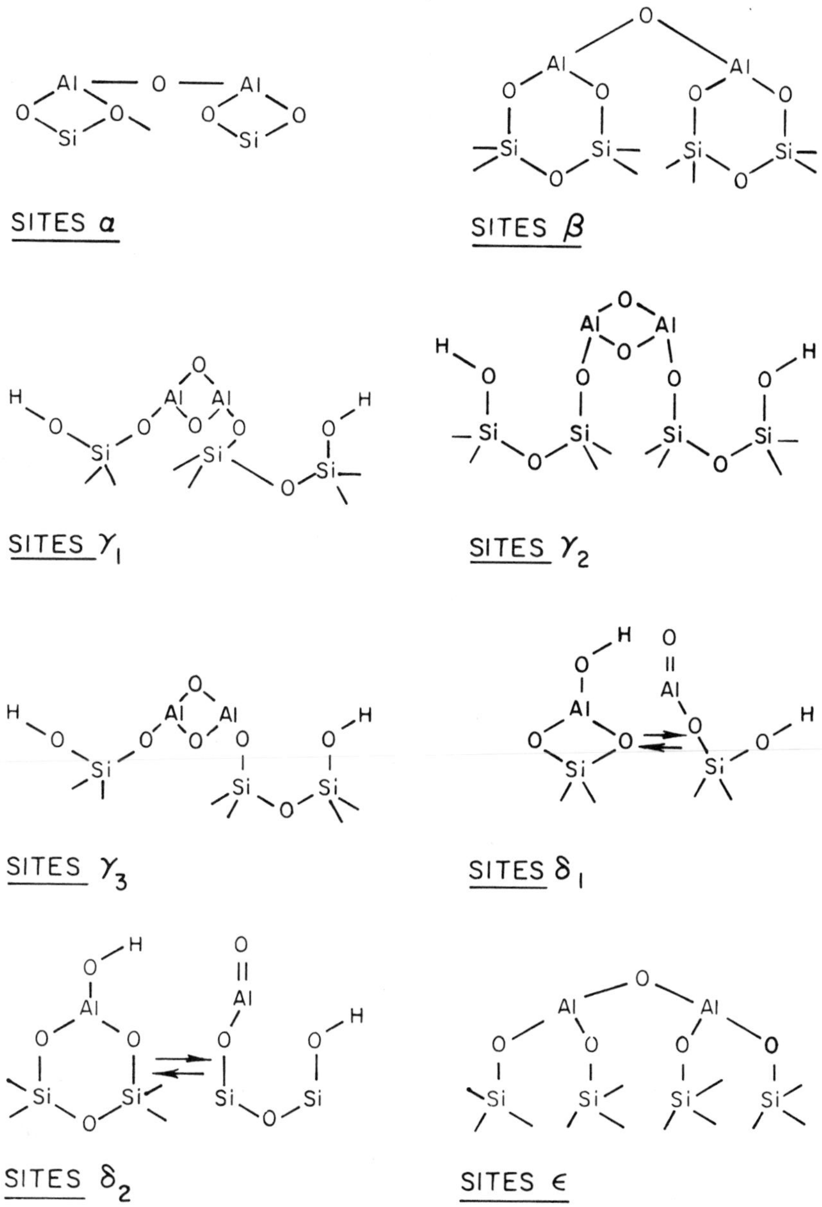

Figure 3.1 Possible types of sites on silica alumina. (From Ref. 17.)

Figure 3-1. Peri's proposal clearly illustrates that there exist a variety of types of bondings surrounding an aluminum atom on the surface. This in turn suggests immediately that each of those types of sites will have specific characteristics of acid strength, steric surroundings, and other properties which will influence its activity as an active site in catalysis. Thus amorphous silica aluminas can be expected to exhibit a wide distribution of active-site acidities, as well as a range of pore sizes and chemical compositions. These properties make it possible to engineer amorphous cracking catalysts into desirable forms. At the same time, the ill-defined nature of the overall structure and of the surfaces in these materials makes any discussion of the active sites on them at best vague. In order to reduce the number of variables involved, we turn to the better defined properties of crystalline aluminosilicates.

3.3 CRYSTALLINE ALUMINOSILICATE CATALYSTS

During the past 20 years a group of crystalline aluminosilicates has attracted wide interest among scientists and catalyst manufacturers. These materials, called zeolites, represent a series of solids with properties that are highly desirable in catalyst formulations [18]. They have high adsorption capacities, high surface areas, and can be made to contain acidic sites of varying strengths. In crystalline aluminosilicates, all aluminum and silicon atoms form tetrahedra which are linked by shared oxygen atoms. A general formula for the composition of such a zeolite can be written as follows:

$$M_x D_{y/2} \cdot Al_m Si_n O_{2(m+n)} \cdot pH_2O \qquad \text{where } x + y = m$$

Forcing aluminum atoms into tetrahedral configurations causes negative charges to appear on each atom. M and D represent mono- and divalent cations which balance such charges in the crystalline structure [19]. These labile cations can be exchanged by contacting the solid zeolite with solutions of other appropriate cations. The ion exchanges in turn result in changes in the acidity of the active sites on the material.

Some 40 natural zeolites are known; several exist in nature in substantial quantities and purities [20], and more than 150 synthetic types have been reported [21,22]. Only a small fraction of all these zeolites are currently of commercial interest, as shown

Table 3.1   Commercial Zeolites and Their Uses

Natural zeolite	Uses	Synthetic zeolite	Uses
Mordenite	⎫	A	Adsorption
Chabazite	⎬ Catalysis	X	Cracking
Erionite	⎬ and water purification	Y	Cracking
Clinoptilolite	⎭	L	Adsorption
		ZSM-5	⎧ MTG process ⎨ Isomerization ⎩ Dewaxing

Source: Data from Ref. 23.

in Table 3.1. Many more will be synthesized and used in catalysis in the future.

Of the commercially interesting materials listed in Table 3.1, only X and Y zeolites are presently used as commercial cracking catalysts. We will therefore concentrate our subsequent discussion on these two zeolites.

The X and Y zeolites are represented by the formula

$$Na_p Al_p Si_{192-p} O_{384} \cdot gH_2O$$

where p ranges from 96 to 74 for X and 74 to 48 for Y, while g goes from about 270 to 250 as aluminum content decreases. In both zeolites the primary building blocks are the silica and alumina tetrahedra, which are arranged at the vertices of a truncated octahedron. The truncated octahedron, known as a sodalite cage, contains 8 hexagonal faces, 6 square faces, 24 vertices, and 36 edges. The sodalite cages are the secondary building blocks in the zeolite. In the next stage of construction, four sodalite cages are arranged in a tetrahedral configuration around a fifth sodalite cage. The five units are joined by hexagonal prisms [24] as shown in Figure 3.2 and constitute the tertiary and characteristic building block of the X and Y zeolites.

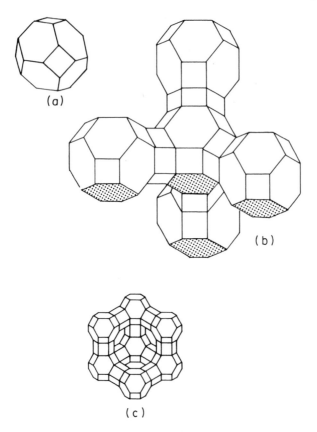

Figure 3.2   Building blocks of faujasite-type zeolites.

The joining of many such building blocks into a regular array
results in a crystalline material whose structure contains large
pores whose smallest constriction is a 12-membered ring of SiO/
AlO entities 9 Å in diameter, shown in Figure 3.3.   This large-
pore network is accessible to a variety of molecules of the type
present in petroleum.

As mentioned before, the negative charges on the tetrahedrally
coordinated aluminums are balanced by cations which are present
in the vicinity of each aluminum atom.   The existence of these
exchangeable cations is the key to obtaining zeolites with enough
acidity to be used as cracking catalysts, as we will soon see.   It
is therefore interesting to know precisely where in the zeolite

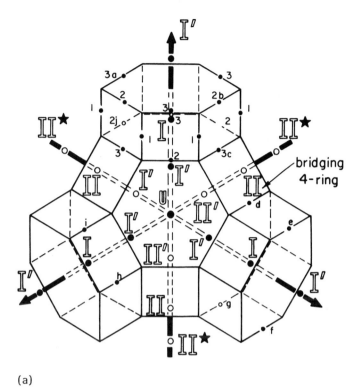

(a)

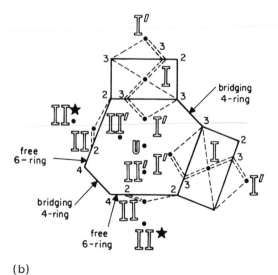

(b)

Figure 3.3 Idealized projection (a) and section (b) through a sodalite unit of faujasite. (From Ref. 24.)

structure these active sites will be located. The literature on the location of the cations in the pore system of the various zeolites has been summarized by Mortier [25].

In the case of sodium X and Y zeolites, five types of cation sites have been defined [26]. There are 16 type I sites situated at the centers of the four hexagonal prisms in a unit cell of the zeolite, as shown in Figure 3.3. Type I' sites are located inside the sodalite cage on the other side of the hexagonal face from type I sites. There are 16 of those sites as well. Type II sites are located on the unjoined hexagonal faces of the sodalite cages and there are 32 of these sites. Thus there are altogether 64 sites associated with the hexagonal faces of the sodalite cell. The remainder are type III sites, which are outside the hexagonal prisms and the sodalite cages and in the large-pore network. These are the sites that are most readily accessible to feed molecules. Not all the sites are filled in each zeolite.

This situation pertains to hydrated zeolites and changes with degrees of hydration and with the type of cation present in the structure. Table 3.2 and Figure 3.3 illustrate how various cations are distributed in hydrated and dehydrated zeolites.

The sodium ions present in the X and Y zeolites, when they are first made, can readily be exchanged by the other mono-, di-, and trivalent cations shown in Table 3.2. The exchange process is very fast and is therefore controlled by diffusion. It is found [27,28] that the first 85% of the sodium is easily exchanged in practice. These are the sodiums which are thought to be located in the supercavity or inside the sodalite cage. The remaining sodiums are harder to remove since, in order to be exchanged, they have to enter the supercage from the interconnecting hexagonal prisms through a relatively narrow window of 2.2 Å. To complete the exchange, it is usually necessary to calcine the material to dehydrate the cations in the hexagonal prisms and thus to increase their mobility inside the zeolite. Such treatment, followed by subsequent exchanges, allows the removal of all the sodium from the starting material.

Of the cations that can be substituted for sodium, the proton is one that is usually not introduced directly. The reason for this is that the framework of the zeolite is sensitive to low-pH solutions. At low pH, aluminum ions are removed from the crystalline framework and the structure collapses, leaving an amorphous material. The critical pH is in the range 3.5 to 4 for zeolite X and 2.5 to 3.0 for zeolite Y [29]. For this reason the exchange of sodium by protons is carried out in an indirect way. In the first step the sodium ions are exchanged by NH_4^+ ions. The

Table 3.2   Distribution of Cation Sites in Zeolites

Zeolite	Site population				
	I	I'	II	II'	III
Hydrated form					
KY	1.3 K	13.3 K	20 K		
NaX	9 Na	8 Na	24 Na	26 Ox	Several sites
KX	9 K	7 K	23 K		
CaX		20 Ca		40 Ox	
La_{29}-X		30 La		32 Ox	
Na_1Ce_{26}-X	11 Ox	9 Ce	21 Ce	> 32 Ox	
Dehydrated form					
KY	5.4 K	18 K	27 K		
NaY	7.8 Na	20.2 Na	31.2 Na		
CaX	7.5 Ca	17 Ca	17 Ca	9 Ca	
Na_1Ce_{26}-X	3.2 Ce	24 Ce		6 Ox	
$Na_{14}H_{42}$-Y	4 Na	3 Na	10 Na		

Source: Ref. 26.

exchange is carried out in steps interspersed by calcination treatments [30]. When all the sodium has been removed, the zeolite is calcined at 500°C to decompose the ammonium ions, leaving behind protons in place of the original sodium.

3.3.1   Synthesis of X and Y Zeolites

The synthesis of X and Y zeolites is well documented in a series of patents, publications, and reviews [31−42]. In principle the zeolites are prepared from alumino-silicate gels, which in turn are prepared from aqueous solutions of sodium aluminate, sodium silicate, and sodium hydroxide. Gel structure depends on such factors as chemical composition and the molecular weight distribution of the starting species in the silicate solution. Once formed, the gels are crystallized at temperatures ranging from 60 to 250°C at atmospheric or autogeneous pressures [35]. The following represents a typical series of requisites [43]:

1. A freshly co-precipated gel

2. A high pH introduced in the form of alkaline methyl hydroxide or other strong base

3. Moderate-temperature hydrothermal conditions at saturated water vapor pressure

4. A high degree of supersaturation of components

The general reaction scheme describing the resultant processes is as follows:

$$NaOH(aq) + \begin{cases} Al(OH)_4Na(aq) \\ \\ Al(OH_3)_3(aq) \end{cases} + SiO_3Na_2(aq) \xrightarrow{\text{room temp}}$$

$$[Na_a(AlO_2)_b(SiO_2)_c \cdot NaOH \cdot H_2O](gel) \xrightarrow{60-250°C}$$

$$Na_p[(AlO_2)_p(SiO_2)_g] \cdot hH_2O (crystals\ in\ suspension)$$

The crystallization takes place in alkali medium at moderate temperatures. The composition of the crystallizing mixture (i.e.,

Table 3.3 Influence of Reactant Composition on the Synthesis of Zeolites

Reaction mixture mole ratio	Primary influence on:
SiO_2/Al_2O_3	Framework composition
H_2O/SiO_2	Rate, crystallization mechanism
$OH^{\ominus}/SiO_2$	Silicate molecular weight
$Na^{\oplus}/SiO_2$	Structure, cation distribution
$R_4N^{\oplus}/SiO_2$	Framework, aluminum content

Source: Ref. 37.

SiO_2, Al_2O_3, $OH^{\ominus}$, H_2O, and cation species) plays a dominant role in determining the type of zeolite that will be produced. These influences are summarized in Table 3.3.

Almost any source of silica and alumina can be used, although the product obtained may depend on the source. The most commonly used starting materials are Na_2SiO_3, silica gel, silica sol, $NaAlO_2$, aluminum sulfate, and various clays. The cation present has a major effect on the structure of the zeolite, and various zeolite types can be obtained from the same initial composition simply by changing the nature of the cation. For instance, on can produce zeolite Y or zeolite L from the same mixture, depending on whether the cation present is $Na^{\oplus}$ or $K^{\oplus}$. To make things more interesting, organic cations can be used to produce a great variety of zeolites of the Z type [44–47].

The degree of crystallinity of the product is judged by comparison with a standard sample of carefully prepared zeolite, using x-ray diffraction, surface area measurements, and ion exchange capacity, as well as electron microscopy. Crystal growth is highly dependent on the SiO_2/Al_2O_3 ratio and on such factors as seeding, low-temperature aging [48,49] of the precursor gel, and reactant purity. Normally, crystallization continues until the aluminum in the mixture is exhausted.

The seeding of mixtures to enhance crystallization brings up an interesting point. To make Y zeolite, crystals of X zeolite are used [37]. This induces radial heterogeneity in the crystals

formed on the seed and points out the fact that such heterogen-
ity may well be part of every crystal; or the various crystals in
a given batch may vary in composition, as they are formed in a
crystallizing mix whose composition changes with the extent of reaction [48−55]. In terms of the influence that such compositional
variations may have, we can only speculate that if specific crystal
compositions are desirable, then homogeneity of the zeolite at this
desired composition can be expected to improve its performance as
a catalyst.

Another aspect of crystal growth is the size of the crystals produced. Crystalline zeolites have pores whose dimensions are of
the order of molecular diameters. This by itself will induce varying degrees of diffusional limitation on the access of feed molecules
into the interior of the crystallite, which contains the vast majority of the surface area of a zeolite. Such effects will range from
negligible to total exclusion (sieving), depending on the size of
the reactant molecule. There is reason to believe that, in this
way, not only is the zeolite sieving the feed with respect to reactant molecule access, but it may also produce a "cage effect"
which traps molecules that do enter, and forces them to undergo
repeated reactions before they can exit. Thus crystallite size,
which is largely ignored in catalyst formulation, may have an important influence on activity and on selectivity.

3.4 THERMAL STABILITY OF CRACKING CATALYSTS

The freshly prepared cracking catalyst is well known to have an
activity vastly different from that of a used catalyst. In fact, it
would be impossible to operate commercial units using fresh catalysts. Thus the stabilization of catalyst before introduction into
the reactor as well as the decay of activity in the reactor are
part of catalyst preparation, in the same sense as ion exchange and
other treatments that have been mentioned above. Before we consider thermal and hydrothermal stabilization of catalysts, we should
remind ourselves of the conditions that the catalyst will encounter
in use.

In a commercial reactor the catalyst will find itself in the presence of hydrocarbons of varying molecular weight and composition,
including components containing sulfur, nitrogen, and heavy metal
atoms. Temperatures in the reactor will typically be between 500
and 600°C. During the cracking process carbonaceous deposits
known as coke are deposited on the catalyst surface, and these

in turn have to be burned off in a separate vessel called a regenerator. Transport of the catalyst from the reactor to the regenerator involves steam stripping of hydrocarbons from the catalyst at about reaction temperature, followed by the exposure of
the catalyst to air and steam in the regenerator. In the regenerator the carbonaceous deposit is burned off in a controlled
manner at temperatures that are maintained below 800°C. The
regenerated catalyst is returned to the reactor and used again
in the cracking process. Thus each cycle involves abrasion and
exposure to hydrocarbons, to heavy atoms present in the feed,
and to steam, air, and high temperatures. The average catalyst
particle undergoes about 15,000 cycles before it is discarded. All
this has an effect on both the physical and chemical properties of
the catalyst in the reactor. We begin by considering the effects
of temperature on the catalysts.

The calcining of amorphous silica alumina in an anhydrous atmosphere up to 800°C results in only a small reduction in pore
volume and a comparably small decrease in surface area. However,
if the heat treatment is carried out in the presence of steam, even
at temperatures as low as 600°C, there is a loss of surface area,
a considerable increase in average pore radius, and a broadening
in the distribution of pore radia [4]. Clearly, in an amorphous
catalyst this physical change alone will result in the loss of some
active sites and may be expected to change the activity distibution
of the remaining sites. The reduction in surface area continues
with time of exposure of the catalyst to the conditions mentioned
above, leading eventually to a fused, nonporous, essentially inactive material [56].

When such a calcining process is carried out in the absence of
steam, the early stages of the calcination show a behavior similar
to that observed in the presence of steam [57]. The reason for
this is that hydroxyl groups are lost from the surface of the raw
catalyst, resulting in the presence of some steam vapor in the
initial stages of an anhydrous calcination. The loss of such hydroxyl sites leads to a loss of activity, since it is the hydroxyls
that supply the acid hydrogen for catalytic cracking. These
methods of stabilization of a fresh catalyst before use in a reactor can involve treatment of some 200 hr at 850°C in order to
achieve the stable properties of an equilibrium catalyst [57].

It is known that certain compositions of amorphous silica alumina are more resistant than others to this type of deactivation.
Partly for this reason, high-alumina amorphous catalysts containing 30% and more alumina were developed. Table 3.4 summarizes
the stability of amorphous silica-alumina catalysts under various

Table 3.4 Influence of Steam Deactivation on the Properties of Si-Al

Steam treatment at 565°C (24 h)	Catalyst					
	24% Al_2O_3		14.3% Al_2O_3		11% Al_2O_3	
	Activity	Surface area (m^2/g)	Activity	Surface area (m^2/g)	Activity	Surface area (m^2/g)
None	116	493	100	520	96	594
1 atm	83	346	48	219	45	291
2 atm	68	272	34	157,	—	—

Source: Ref. 4.

conditions of deactivation as evidenced by their activity for cracking a Texas gas oil.

The physical changes described above are accompanied by chemical changes on the surface of the catalyst. On the surface of an
amorphous silica alumina there are two types of acid sites,
Brønsted and Lewis, as illustrated below. The Lewis sites are
seen to be the consequence of the dehydration of Brønsted sites.

Bronsted site   Lewis site

These sites have been characterized by a number of physical
and chemical methods, which will be discussed later. The two
types of sites have differing effects on the feedstock, and the
process of sintering or stabilization results in a change not only
in their absolute number but also in their relative amounts. When
silica alumina is heated at temperatures above 450°C, the total
number of Brønsted sites begins to decrease; most are destroyed
by the time a sample is heated to 700°C in the absence of steam.
During this treatment water is generated and the number of Lewis
sites increased [3]. Fortunately, the process is almost completely
reversible by the addition of water to the dehydroxylated catalysts if the dehydroxylation temperature does not exceed about
600°C. A similar treatment in the presence of steam also affects
the total number of acid sites, but in this case it is the strongest acid sites that are preferentially removed [58].

Zeolite catalysts, although similar in composition to the silica
alumina, are much more resistant to heat and steam treatment
than are the amorphous catalysts [59–61]. Structural collapse
does not occur in zeolite catalysts until much higher temperatures,
such as 1100°C, are reached. It is generally accepted that the
main reason for this increased stability is the geometric structure
of the crystalline framework composing the zeolite itself. Several
other factors are also involved, including the nature of the exchangeable cation, the level of the exchange, and the silica-to-
alumina ratio. Figure 3.4 illustrates the relative stability of
zeolite catalysts as the silica-to-alumina ratio is changed.

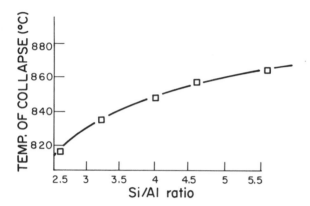

Figure 3.4 Influence of the Si/Al ratio on the thermal stability of a faujasite-type zeolite. (From Ref. 62.)

The nature of the exchangeable cations has a profound influence on the stability of zeolites. The temperature at which the crystalline structure collapses increases with the size of the cations in the alkaline series. This is thought to be due [63] to the relative ability of the various cations to fill voids in the crystal after dehydration. Divalent cations have a still higher stabilizing effect, while trivalent ions lead to the most stable zeolites [64]. In commercial catalysts the sodium level of the zeolite is kept as low as possible, to prevent structural collapse of the zeolite under conditions encountered in the reactor [65].

As was the case with amorphous silica alumina, zeolites are much less stable in the presence of steam. In extreme conditions the steaming of crystalline materials leads to total collapse and a resulting amorphous mass. Under other circumstances the crystalline structure rearranges in the presence of steam to a more stable form than that which was present originally.

The ion exchange procedure is also important in controlling the stability of zeolite catalysts. McDaniel and Maher [30,66,67] have obtained a very stable type of Y zeolite in which a high-temperature stabilization step is required. The resulting product presents exceptional stability in both high-temperature treatment and steaming. Such materials are usually hydrogen Y zeolites and fall in the category of so-called "ultrastable" materials [68,69]. It is now known that the ultrastable structures are characterized by a contraction of between 1 and 1.5% in the unit cell dimensions and that this contraction is related to the removal of aluminum atoms from the framework of the crystal and to the location of these aluminums as cations in exchange positions in the cages. Kerr, for instance [70,71], has prepared an ultrastable Y zeolite by extracting aluminum from the framework using EDTA and calcining the resultant aluminum-deficient zeolite at 800°C. A maximum effect was obtained upon removal of some 30% of the aluminum.

Each treatment of a zeolite is important, not only from the structural point of view, but also from the point of view of the quality and quantity of the active sites. When an HY zeolite is heated at temperatures above 500°C it loses hydroxyl groups attached to the aluminum atoms, with the resultant formation of tricoordinated aluminum and the elimination of water, as follows:

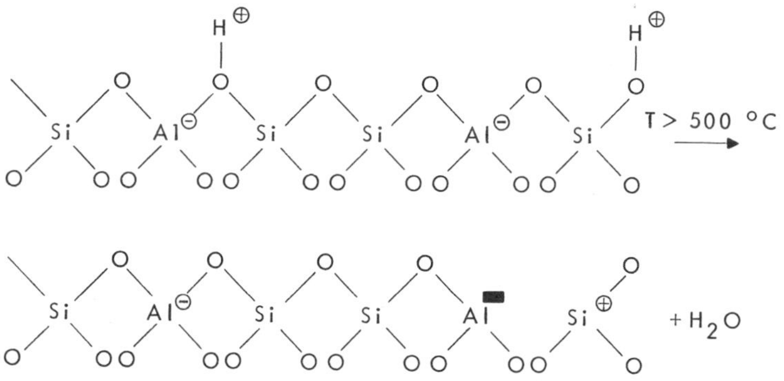

As in the case of amorphous catalysts, such chemical changes result in significant changes in the number, distribution, and nature of the acid groups present on the catalyst. In general, it is found that the most acidic sites are removed first by the thermal treatment of zeolites [72]. In all such treatments significant changes in activity and selectivity are produced, complicating the evaluation of a catalyst by introducing complex time-dependent effects into the consideration of its practical application in commercial reactors.

3.5   SURFACE ACIDITY OF CRACKING CATALYSTS

Several methods have been evolved for measuring the nature and number of acid sites on the surfaces of solid acids; these have been reviewed thoroughly by a number of authors [3,73,74].

Methods have been developed that distinguish between Brønsted
and Lewis acid sites, while others measure the total amount and
distribution of the various acids present. Unfortunately, most of
these methods are applied at temperatures far removed from those
actually used in catalytic cracking and can therefore only indicate
certain trends rather than measure phenomena at reaction conditions. As a result, the significance of acidity measurements on
catalysts must usually be interpreted using more direct evidence
of activity-related phenomena such as model reactions. The cracking of normal hexane and alkyl aromatics, isomerization of xylenes,
and alcohol dehydration are examples of reactions used to characterize the activity of cracking catalysts.

On the basis of such investigations it has been found that only
a small fraction of the total active sites present on a catalyst surface are in fact engaged in any given test reaction [75]. Furthermore, each of these reactions seems to occur on a different set of
active sites, presumably differing from other active sites by their
acid strength. If this picture is in fact correct, it is clear that
a large number of reactions requiring acid sites of various
strengths are needed to map the total activity of the catalyst.
Such a method of quantification of acid-site strength distribution
would be far too laborious to be used routinely. One is therefore forced to rely on the evidence presented by various titera-
tions and spectroscopic methods as the major source of information
on catalytic acid sites. The following list of the various character-
ization methods used in practice is taken from various sources
[73,76−89].

Direct Determination of Acidity

1. Methods measuring the total number of acid sites

    a. Amine titration

    b. Adsorption-desorption of gaseous bases

    c. Aqueous titration

    d. Calorimetric methods

    e. Poisoning by bases of various strengths

2. Methods of determining the acid strength distribution

    a. Adsorption of color indicators

    b. Spectral photometry

    c. Adsorption-desorption of gaseous bases

d. Calorimetry

e. Benzene adsorption

f. The shifting of IR absorption bands

g. Proton mobility by NMR

3. Methods of determining the nature of acid sites

   a. Brønsted sites

     (1) Cation exchange

     (2) Titration: pyridine, 2,6-dimethylpyridine, NH_3

     (3) Reaction with hydrides

     (4) Spectroscopy: optical, ESR

   b. Lewis sites

     (1) Reaction with electron-donating reactants

     (2) Titration: pyridine

     (3) Spectroscopy: optical

Indirect Determination of Acid Character Through Chemical Reaction

1. Conversion of sucrose

2. Esterification of phthalic acid

3. Dealkylation of aromatics

4. Decomposition of formic acid

5. Isomerization of hydrocarbons

6. Dehydration of alcohols

7. Depolymerization of propyl aldehyde

8. Disproportionation of halogenated hydrocarbons

9. *n*-Hexane cracking

10. H_2-D_2 exchange

3.5.1 Acid Sites on Silica-Alumina Cracking Catalysts

In general, it has been established that amorphous silica alumina contains both Brønsted and Lewis acid sites and that both these

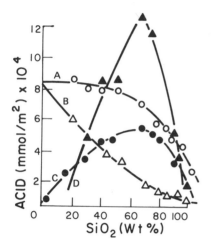

Figure 3.5   Influence of the Si/Al ratio on the acidity of a silica-alumina catalyst: A, total acid $H_0 \geqslant 1.5$; B, Lewis acid; C, Brønsted acid (by difference); D, Brønsted acid by NH_4 exchange. (From Ref. 91.)

types of sites appear to be involved in the variety of processes that occur in catalytic cracking [90].   Figure 3.5 shows the distribution of Brønsted and Lewis acid sites found by Shiba et al. [91] on a sample of amorphous silica alumina prepared by mixing alumina gel and silica gel and heating the mixture at 500°C before measuring the acidity.

From this figure we see that the various methods of measuring catalyst acidity agree, at best, qualitatively, but differ to a substantial extent in their quantitative evaluation of acid sites.   Second, a maximum in Brønsted acidity is found at a silica content of 70%, corresponding to the 30% alumina catalyst which was mentioned earlier as the preferred composition for an amorphous cracking catalyst.

Lewis and Brønsted active sites can be studied using the IR spectra of pyridine adsorbed on the catalyst surface [92–94]. Figure 3.6 shows an IR spectrum of pyridine adsorbed on a 25% silica-alumina cracking catalyst.

There we see that the Brønsted acid sites interact with pyridine to form a pyridinium ion whose principal characteristic infrared (IR) band appears at 1548 cm^{-1}, while pyridine coordinates with Lewis acid sites, leading to a characteristic absorption band at 1457 cm^{-1}.

By heating a sample containing adsorbed pyridine, one can observe how desorption proceeds as a function of sample temperature. It is inferred that the strongest sites are the ones that desorb at the highest temperature, and thus a rough quantitative evaluation of acid-site strength distribution can be obtained [95]. In this way, using IR spectroscopy and temperature-programmed desorption of pyridine, Schwarz et al. [96] estimate that there are 0.04 μmol of Brønsted acid sites per square meter (2.4×10^{12} sites/cm^2) and 0.23 μmol of Lewis acid sites per square meter (1.3×10^{13} sites/cm^2) on the surface of a 10 wt % silica alumina. Other studies [97–99] show that there are 5×10^{12} strong Lewis sites per square centimeter on a similar cracking catalyst.

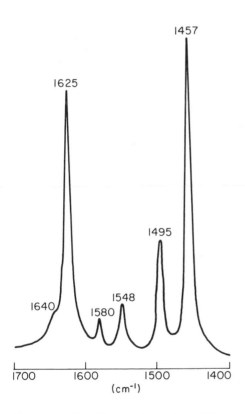

Figure 3.6   IR spectra of pyridine adsorbed on a 25% silica-alumina cracking catalyst.   Coordinately bonded pyridine on Lewis sites at 1625, 1580, 1495, and 1457 cm^{-1}; pyridinium ion on Brønsted sites at 1640, 1625, 1548, and 1495 cm^{-1}.

The number of hydroxyl groups obtained from absorbence measurements of IR spectra in a silica alumina which has been dried at 600°C is reported to be 8.3×10^{13} OH per square centimeter [100]. Other authors report 2.4×10^{12} OH per square centimeter on investigating π complexes with adsorbed cumeme [101]. Using catalytic techniques, several authors have found a Brønsted site density ranging from 2×10^6 to 9×10^7 sites per square centimeter, depending on the type of silica alumina used [102−104]. Clearly, the numbers obtained from various absorbence measurements differ from those obtained by cumene π complex formation, which in turn differ by a very large factor from site densities estimated by catalytic techniques.

It has been shown that cumene cracking and the accompanying reactions, such as disproportionation, chain isomerization, and so on, can be used for comparing not only the number of Brønsted acid sites but also the number of Lewis acid sites on various catalysts. Using these reactions, Marczewski and Wojciechowski [105] have compared two silica aluminas, one with 14%, the other one with 25% alumina. By comparing initial rate constants and activation energies for all the reactions involved, they conclude that the differences in the activity of the two catalysts is related strictly to the total number of acid sites and to the ratio of Brønsted to Lewis sites. Furthermore, it seems that 25% amorphous silica alumina is some four times more active than is 14% silica alumina, while the ratio of Brønsted to Lewis sites is the same for both catalysts. These results indicate that the proportion of active acid sites can be measured by catalytic techniques to reveal those active for any given reaction. The number of acid sites obtained by acid-base interaction can only be taken as a rough indicator of the situation that will influence catalysis. Furthermore, it is not enough to measure the absolute number of acid sites; rather, it is necessary to obtain the acid-site strength distribution in order to correlate activity for a given reaction with the number of acid sites present.

Butyl amine titration is by far the most common method used for the study of acid-site strength distribution on catalysts. This method should, however, be used with considerable care since there is much room for misinterpretation [106,107]. Using Benesi's method, the results depend on factors such as the amount of indicator, the particle size of the catalyst, the concentration of the titrant, and the time allowed for adsorption to reach equilibrium. In the case of amorphous silica-alumina and of zeolites, this method gives a higher number of strong acid sites than do other techniques, such as microcalorimetry, ammonia desorption, and gas chromatography.

An alternative method proposed by Deba and Hall [107] consists of measuring the strength by chemisorption of bases of different pK_a values. By using ultraviolet spectrophotometric methods and following n-butylamine titration, Take et al. [108] have studied the acid strength distribution on several silica aluminas. They report that the strongest acid sites on these catalysts had HO values of -12.8.

3.5.2 Acid Sites on Zeolite Cracking Catalysts

The forms of zeolites containing alkaline metal ions are inactive in the cracking reaction. At the same time it is found that there is a complete absence of strong acid sites on the surfaces of such materials. Therefore, the preparation of an active cracking catalyst from a zeolite requires the formation of strong acid sites. This is done by exchanging varying amounts of the originally present sodium ions by ammonium ions or by di- or trivalent cations.

The most obvious method of generating acidity on the surface of the zeolite is the replacement of sodium ions by protons. This is normally done by the exchange of sodium with ammonia, since direct exchange with an acid destroys the crystalline framework of the zeolite. Once the ammonia is placed inside the cages and the sodium is removed, the catalyst is calcined at a higher temperature, decomposing ammonium ions, driving off ammonia, and leaving protons in the place of the original sodium ions [107−112]. Using this procedure, sites strong enough to form H_3O^{+} and $H_5O_2^{+}$ can be created [113].

The properties of the final catalyst are strongly dependent on the final activation step, the time of decomposition, the temperature, and the atmosphere surrounding the catalyst at the time when the ammonium ions are decomposed.

Uytterhoeven et al. [110] have studied samples of NH_4NaX and NH_4NaY zeolite and report that when these materials are calcined at a temperature of 300°C, IR bands at 3745, 3640, and 3540-cm^{-1} wave numbers are formed. These bands correspond to the stretching frequencies of acidic OH groups which are able to react with gaseous NH$_3$. They report that the intensity of these bands decreased when the temperature of the treatment was increased and that deamination and dehydroxylation occur in the same temperature range [114]. In general, the bands are found to diminish when an organic base is absorbed on the catalyst, thus proving that they correspond to acidic OH groups. In another study, Ward and Hansford [115] have shown that the 3640-cm^{-1} band increases almost linearly with the level of exchange, whereas the 3540-cm^{-1} band does not increase until 60% of the intial sodium has been replaced. Thus it seems that various types of acid sites appear at various levels of exchange.

At the same time, several authors report that reactions catalyzed by Brønsted sites show a nonlinear increase in activity with the level of exchange. It has been found that at exchange levels below 70%, the catalytic cracking activity of a zeolite is minimal. Above this level of exchange, activity increases sharply for a number of the more difficult carbenium ion reactions [115–119]. Thus it is the sites which are formed at high levels of exchange that seem to have the necessary acidity for catalytic cracking.

The amount of activity induced in an HY catalyst depends somewhat on the method of preparation. We have already seen that calcination of zeolite in the presence of NH$_3$ or H$_2$O vapor results in the production of ultrastable zeolites. There are big differences between HY and ultrastable HY zeolites, not only in their stability, but also in the quantity and quality of acidity as shown by their IR spectra [120–123].

Inspection of Figure 3.7 shows that the ultrastable zeolite has fewer hydroxyl groups at 3650- and 3540-cm^{-1} wave numbers, both of which are believed to be strongly acidic. The band at 3740-cm^{-1} wave numbers is much stronger than in the decationized zeolite. This band has been associated with nonacidic silanol groups. The two bands at 3675 and 3600-cm^{-1} correspond to sites that do not interact with pyridine and are thought to be either nonacidic or not accessible to large molecules [122]. In general, therefore, it can be said that the formation of ultrastable zeolites decreases the total number of acid sites and tends to remove the strongest sites.

As in amorphous silica alumina, when zeolite catalysts are heated above 400°C, the number of Brønsted sites decreases and the number of Lewis sites increases [75]. Above 800°C the zeolite has only Lewis sites remaining. It is found that the sum of the Brønsted sites plus twice the number of Lewis sites is constant

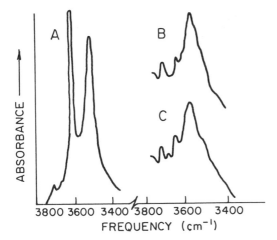

Figure 3.7 IR spectra of HNaY zeolites: A, decationized; B, McDaniel and Maher's ultrastable; C, Kerr's ultrastable. (From Ref. 120.)

throughout this range of temperatures [75]. If calcination is restricted to below 600°C, most of the Lewis acid sites can be rehydrated by the addition of water to restore Brønsted sites [75]. Furthermore, Eberly reports that the strength of the Lewis sites formed in this way is higher than that of the Brønsted sites [124].

Another factor affecting the total acidity as well as the strength distribution in zeolites is the silica-to-alumina ratio [75,125−130]. It seems that the acid strength in a homologous series of faujasites increases as the number of aluminum atoms decreases [131]. The silica-to-alumina ratio can in turn be varied over a wide range by changes in the composition of the initial synthesis mixture or by dealumination of a solid zeolite [132−135]. From such work it was found that X zeolite has fewer and weaker sites than Y [136]. This seems rather surprising at first, since X has higher exchange capacity due to a higher alumina content, and consequently a higher number of protons can, theoretically, be developed on its surface.

To further complicate the matter, it was shown [136] that in dealuminzed Y zeolites which had been extracted with EDTA, the extraction of some 33% of all the alumina initially present does not affect the activity of the catalyst in isooctane cracking. Activity decreases if further extraction is carried out. Since only the strongest acid sites are believed to be active in this reaction, the authors conclude that the 33% of aluminum extracted does not correspond to the strongly acid sites. This also seems surprising, since there are no a priori distinctions among the various lattice alumina in the zeolite. In fact, it suggest that extraction is governed by local acidity at the various aluminum atoms [136–138] and that such local acidity is governed by factors that are not yet fully understood but may be connected with crystal heterogeneity.

In the case of ZSM-5, a linear relationship is observed between catalytic activity in *n*-hexane cracking and aluminum content in the range 10 to 10,000 ppm of aluminum [139,140] without any maximum. On the other hand, in faujasites a maximum in strong acid sites and cracking activity occurs at Si/Al ratios between 4 and 8 [141,142]. Thus the maximum in acidity is a function of the zeolite structure and the surroundings of individual aluminum atoms in the lattice.

In both amorphous and crystalline silica aluminas there is generally a Si/Al ratio at which the desired acidity is optimized. The fundamental reasons for this, especially in the regular structures of zeolite, may have something to do with crystal homogeneity (discussed in Section 3.3.1), but this is by no means clear.

3.5.3   Modification of Zeolites by Ion Exchange

A common way of preparing zeolites which are catalytically active is to exchange the sodium ions by di- or trivalent cations and heat the resulting material to temperatures above 300°C. Almost all the ions in the periodic system can be introduced into zeolites in this way [143]. The activity of such "exchanged" zeolites was first explained by associating assumed electrostatic fields near the cations with the formation of carbocations by the polarization of the CH bond of the reactant molecule [144]. This hypothesis has by now largely been abandoned, and at present the polyvalent cation forms of zeolites are thought to form acid sites that are similar or identical in nature to those formed in H zeolites [145]. The formation of such acid sites is thought to take place by the hydrolysis of water associated with the polyvalent cations [146], as follows:

$$Me^{n\oplus}(H_2O)_{n-1} \xrightarrow{\text{heat}} Me(OH)^{\oplus}_{n-1} + (n-1)H^{\oplus}$$

The total number of protonic sites formed will depend on the stichiometry of exchange, the cation exchange capacity of the zeolite, the degree of the exchange, and the structure of the zeolite framework [147]. Their activity governs their ability to form carbocations on the organic feed molecules, and consequently the activity of the catalyst will depend on the number of appropriate acid sites present on the surface. In a number of instances it has been found that the same activation energy for a reaction applies on a variety of zeolites at a variety of exchange levels, with various cations and at different silica-to-alumina ratios [118,148—151]. This in turn indicates that as far as the reaction is concerned, the sites are identical in all such materials. Thus the benefit of exchanging one ion rather than another into the zeolite lies in the resultant stability of the catalyst and in the number of appropriate active sites created. Rare-earth-exchanged zeolites are generally more stable than those exchanged with other ions [151,152] and have been used in industry for this reason. The ultrastable HY forms are even more attractive, both because of their stability and because of their activity for catalytic cracking.

From the point of view of catalysts, three factors concerning the active sites are important: the number of sites per unit weight of catalyst, the ratio of Brønsted to Lewis sites, and the acid strength distribution of each type of site. In principle the number of Brønsted sites should correspond to the number of aluminum atoms in the zeolite, but in practice many factors, including the degree of crystallinity, level of exchange, and cation type, conspire to make this value unpredictable from first principles.

The ratio of Brønsted to Lewis sites is generally a function of the pretreatment conditions for a given zeolite type. In general, heating the catalyst to a high temperature, say about 700°C, results in a loss of Brønsted acidity with a corresponding increase in Lewis acidity. This phenomenon is accompanied by the elimination of water, and if the thermal treatment is not too extreme, the Brønsted sites can be regenerated at lower temperatures by the addition of water. In practical terms this means that the ratio of Brønsted to Lewis sites, and all the catalytic consequences of this ratio, will to some extent be a function of the specific operating conditions in the reactor.

The acid strength of the sites on the catalyst depends on all of the following: Si/Al ratio, zeolite type, cation type, and pretreatment conditions. It seems that the above can be understood as contributing to two main effects: short range and long range [153]. Of these, the short-range effect is due to the nearest charge-carrying neighbors of the aluminum atom. These are no doubt involved in a buffering action provided by their associated cations [154]. The pertinent parameters are therefore distance and the number of cations or hydroxyls that surround the aluminum in question [138,155—158]. The long-range interactions are due to the combined effect of all the aluminum atoms. Thus the density of aluminum atoms, which leads to some average environment for each aluminum atom, also influences the distribution of site acidities [127,129,131,154,159]. When the aluminum content of a given zeolite structure decreases, the distribution of acid strengths shifts in the direction of higher acidities, as evidenced by a decrease in the IR wave number of acid hydroxyls in the structure [160]. In terms of relative strength, Lewis acids are thought to be stronger than Brønsted sites in zeolites [76,161,162].

In view of the great number of variables involved in the formulation of cracking catalysts, many optima can be envisioned in their formulation for specific applications. The preparation of an optimized catalyst requires the production of a hard solid, with the desired composition, the desired crystal structure, and the desired micro- and macroporosity. This must be activated by a suitable ion exchange followed by an appropriate thermal treatment. To complicate matters further, various other functions may be added to the catalyst to assist in regeneration or pollution control without interfering with the cracking function, as we will see in the following section.

3.6 COMMERCIAL CRACKING CATALYSTS

The commercial application of cracking catalysts involves the conversion of highly complex feeds, consisting of high-molecular-weight petroleum distillates, into materials containing aromatics and large portions of C_5 to C_{10} molecules suitable for motor fuel applications. Unavoidably, parallel reactions take place, resulting in the production of light gas and in coke deposition on the catalyst. Thus one of the desirable characteristics of a commercial cracking catalyst is a high selectivity for the C_5-to-C_{10} fraction, low gas and coke yield, and a high yield of aromatics and isomeric alkanes in the gasoline range.

The mention of coke brings up the point that all cracking catalysts lose activity in use by the formation of coke on their surfaces. The coke is both a by-product and a poison of catalyst activity. Since the coking reaction is very fast, catalyst activity decreases rapidly and the catalyst has to be removed for regeneration after a relatively short time on stream. Commercial reactors take this into account by circulating the catalyst between the reactor and a regenerator. Clearly, the required circulation rate will be higher if the catalyst fouls more rapidly. A high circulation rate in turn leads to problems with attrition and consequent catalyst losses in the form of fines. The successful catalyst must therefore have high attrition resistance as well as high gasoline selectivity and stability. Nevertheless, even the best catalysts are fouled by coke in a matter of seconds or minutes and need regeneration. Regeneration is carried out in the presence of air and water vapor, at temperatures higher than those used in cracking. It is here that the thermal and hydrothermal stability discussed before becomes important.

The first cracking catalysts used commercially on a large scale were synthetic amorphous silica aluminas and silica magnesias [163]. These correspond in their properties to the catalysts already described in proceding sections. In the early 1960s, zeolites [164–166] with rare-earth ions replacing the sodium cation were added to the amorphous matrix to improve catalyst activity and selectivity. The new catalysts had remarkably higher activity, much better selectivity for gasoline, and a higher hydrothermal stability than the amorphous catalysts used up to that time. The typical commercial catalyst of this type consists of some 10 to 20% zeolite in an attrition-resistant silica-alumina matrix which provides the bulk of the catalyst mass. The matrix itself is highly porous and allows access to the crystallites of zeolite embedded in the interior of the particle. In practice, the matrix is almost completely inert; in comparison with the activity of the zeolite, the product distribution and activity of commercial zeolite-containing catalysts can be ascribed to the small percentage of zeolite present [167].

The influence of the zeolite on the yield and selectivity of the cracking catalyst has been reviewed by Magee and Blazek [172]. Their results can be summarized in part by noting that an increase in the amount of zeolite present in the catalyst increases the yield of gasoline and light cycle oil, while coke and dry gas make is decreased at any given level of conversion. At the same time the aromatic content and the octane number of the gasoline fraction increase.

Practical cracking catalysts are designed to perform other functions in addition to the cracking function itself. Since the coke formed on the catalyst must be burned off in the regenerator, and since the heat of the cracking reaction is supplied by the hot catalyst returning from the regenerator to the reactor, it is desirable to make just enough coke to supply the heat of reaction. Furthermore, it is desirable to burn the coke on the catalyst to CO_2 rather than CO, since the greater heat of reaction in burning to CO_2 reduces the amount of coke necessary for reactor heat balances. At the same time, the production of CO_2 in the regenerator removes the need for afterburning of CO in the flue gases. Consequently, small amounts of noble metals are added to the cracking catalyst to act as catalysts for the conversion of CO to CO_2 during regeneration [176–178]. This in turn causes the bed temperature in the regenerator to rise and places greater demands on the thermal stability of the catalyst itself.

The SO_x and NO_x which are produced from compounds in the feedstocks that contain sulfur and nitrogen must also be controlled in the emissions from the reactor. The sulfur problem can be solved by a variety of pretreatments of the feed or by adding a new function to the cracking catalyst. If a base metal [179] is introduced into the cracking catalyst, it captures the SO_x in the regenerator, only to release it as H_2S in the reactor. The H_2S can then be easily stripped from the light product gases issuing from the reactor [180]. The sequence of reactions involved in this scheme is as follows:

Sulfur capture in the regenerator:

$$S + O_2 \longrightarrow SO_2$$

$$2SO_2 + O_2 \longrightarrow 2SO_3$$

$$MeO_x + SO_2 \longrightarrow MeO_{(x-2)}SO_4$$

$$MeO_x + SO_3 \longrightarrow MeO_{(x-1)}SO_4$$

Sulfur release in the reactor:

$$MeO_{(x-1)}SO_4 + 4H_2 \longrightarrow MeO_x + H_2S + 3H_2O$$

$$MeO_x + H_2S \rightleftharpoons MeO_{(x-1)}S + H_2O$$

Sulfur release in the stripper:

$$MeO_{(x-1)}S + H_2O \rightleftharpoons MeO_x + H_2S$$

The elimination of NO_x is a more difficult problem and attempts are being made at its solution. The amount of such nitrogen emission lies between 8 and 130 kg per 1000 barrels of a typical FCC feed and can present a serious pollution problem in congested and smog-prone areas of the world [181].

When heavier feedstocks are cracked [182] one also encounters the problems of organometallic compounds containing the heavy metals, Ni, V, and Fe, which tend to deposit rapidly on the external surface of the catalyst. Such deposits lead to a significant increase in the amount of coke formed and in the yield of light gases, while decreasing gasoline yield. The problem can be tackled in a variety of ways. The removal of the outer surface of catalyst particles by attrition removes the deposited metal, but also leads to high and costly catalyst losses. A more economical method is to increase the zeolite content of the catalyst, since zeolites have a higher resistance to metal poisoning than does amorphous silica alumina. One can also demetalize the feed in a preliminary treatment, although this is in general an expensive process in its own right.

Recently, a new and attractive solution to this problem has been introduced. The metals poisoning problem is ameliorated by the addition of metal passivating agents. The passivating agents are organometallic complexes of antimony, bismuth, phosphorus, tin, and related elements [183–189]. Such additives are introduced into the reactor in the form of a soluble compound which is added to the feed when required. Although the mechanism of action of the passivating agents is not clear, it appears that they poison the gas, making catalytic activity of the deposited metals and their efficacy undeniable.

3.6.1   Preparation

Most commercial cracking catalysts contain a synthetic faujasite, which is the catalytically active part of the material, and an amorphous matrix, which is less important catalytically but is responsible for the physical shape and strength of the catalyst [167]. The operations necessary for the preparation of such a catalyst are as follows:

1.   Synthesis of the zeolite

2.   Ion exchange of the zeolite

3.   Activation of the zeolite

4. Synthesis of the matrix gel

5. Combining of the zeolite with matrix gel

6. Washing and exchanging of the wet catalyst

7. Drying and calcination of the catalyst

   Of the above, the preparation of the zeolites has been discussed
in Sections 3.3 to 3.5. Commercial catalysts are generally exchanged with mixed rare-earth ions, $NH_4^{\oplus}$ or magnesium ion, or
mixtures of these.

   The matrices vary from catalyst to catalyst in their Si/Al ratio,
residual sodium level, porosity, and hardness. Usually, 10 to 25%
of zeolite is incorporated in the matrix of a commercial catalyst.
The matrix itself can be made of silica alumina or alumina gel,
clay, or clay plus gel [172].

   The silica-alumina gel is obtained by neutralizing a sodium
silicate solution with sulfuric acid to obtain the silica hydrogel,
which is reacted with aluminum ions to produce the desired silica-
alumina composition. An important property of the final material
is the hardness and pore-size distribution of the catalyst in its
operating form. A matrix with a high proportion of small pores is
hydrothermally unstable and will lose porosity in operation. This
will encapsulate the zeolite particles in the interior of the matrix
and consequently reduce the activity of the catalyst. In the synthesis of the silica-alumina matrix, the final pore structure can be
controlled to some extent during the neutralization step and during
the addition of aluminum ions [7,190—193]. Silica-magnesia matrices have been prepared but are not used in practice [194].

   Other matrices are made from natural clays. When kaolin, hal-
loysite, or bentonite is treated with sulfuric acid, aluminum is extracted from the framework of the clay, leaving behind a leached
silicon-rich clay [168—171]. The leached clay is reacted with sodium silicate in a basic medium and the silicated clay is exchanged
with rare earth or magnesium ions and mixed with an aluminated
acid-treated clay [172]. One type of cracking catalyst is prepared from kaolin by partially converting it to something like Y
zeolite, while the remainder of the kaolin acts as a matrix [173—
175]. Such clay-based catalysts contain a high proportion of
macropores and hence are important in cracking heavy feedstocks
whose average molecular diameters are large and whose reactions
in small pore catalysts are diffusion limited.

   The particle size of the catalyst is usually 40 to 150 μm, while
zeolite crystals are usually about 3 μm. Thus the matrix allows the
production of catalysts in microspheroidal form for use in fluidized-

bed or riser reactors. The microspheroidal form is obtained by
spray drying, either of the gel formed from waterglass and containing the zeolite, or from a sol, with gelation taking place during spray drying. In either case, well-shaped microspheres of the
desired size are readily made [195,172].

The normally encountered variations in matrix composition do
not have a significant effect on the product distribution in cracking. The exception is the silica-magnesia matrix, which enhances
gasoline selectivity and increases the yield of light cycle oils [196].
In general, the role of the matrix is to provide bulk and support
for the zeolite crystals, minimize catalyst loss by attrition, assist
fluidization by providing the desired size and shape of particles,
and provide a sink for poisons present in the feed.

3.6.2   Physical and Chemical Properties

The surface area of the zeolite in a freshly prepared cracking catalyst is in the range 550 to 650 m^2/g, while that of the matrix depends very much on the catalyst and can range from 40 to 350
m^2/g. The general tendency is to use matrices of low surface
area, since such catalysts have a lower selectivity for coke and
are resistant to poisoning by metals.

The pore volume of the catalyst is controlled mainly by the
characteristics of the matrix, in particular by its surface area.
In general, the bigger the pore volume for a given surface area,
the larger the average diameter of the pores. Catalysts with
large pores are easier to regenerate, are better for treating
heavy feeds, and show greater hydrothermal stability in operation.
Commercial catalysts with areas in the range 100 to 400 m^2/g have
pore volumes in the range 0.20 to 0.50 ml/g and an average pore
diameter in the range 50 to 80 Å [198].

In "equilibrium" catalysts, after substantial time on stream, the
differences between the high-area and low-area catalysts are considerably reduced [198]. The remaining differences are in surface
area [198], while pore volume and pore diameter become almost
identical, as shown in Table 3.5.

The rapid decrease in surface area takes place mainly in the
severe hydrothermal conditions of the regeneration. About 20%
of the zeolite and 50% of the matrix are rapidly deactivated at regenerator conditions of about 200 mmHg of steam and 700°C. Subsequent deactivation proceeds at a rate of some 0.5% of activity
per day, making catalyst life fairly long in practice [199]. The
decay rate will be considerably increased by the presence of metal
poisons in the feed.

Table 3.5   Influence of Time on Stream on the Textural
Properties of Cracking Catalysts

	Fresh catalyst		Equilibrium catalyst	
Property	Low area	High area	Low area	High area
Surface area (m^2/g)	100	300	56	110
Pore volume (ml/g)	0.25	0.42	0.27	0.37

Source: Ref. 198.

    The physical strength of the catalyst and in particular its re-
sistance to attrition must be carefully balanced against the cost
of production, the useful life of the catalyst, and the desirability
of abrading the outer surface at a rate that will help to keep
poison metal loading low. The metallic poisons are known to be
concentrated on the surface of the catalyst, and hence abrasion
of the surface is not totally undesirable. As with the other
properties, equilibrium catalysts tend to have similar abrasion
resistance, regardless of their initial physical condition. However, certain factors, such as zeolite content, density, and pore
volume, do influence abrasion resistance and can be used to advantage in achieving specified abrasion rates.
    The particle-size distribution of a fluidized-bed catalyst has a
major influence on catalyst losses and is therefore important in
catalyst formulation. Although the desired distribution is highly
dependent on the reactor design, operating conditions, and sep-
arator efficiencies, in general particles smaller than 20 μm are
rapidly lost from the system, while those between 20 and 40 μm
are only partially recovered. Nevertheless, a certain proportion
of such particles is necessary to promote smooth fluidization of
the bed. The desired size distribution and activity are difficult
to achieve under any but real operating conditions, and as a consequence there is a market in used equilibrium catalysts for start-
ups and other occasions where a major addition of fresh catalyst
will cause intolerable conditions to arise in the reactor.

3.6.3 Performance

In general, the activity of commercial cracking catalysts increases with the zeolite content of the catalyst. Different methods of catalyst activation and matrix formulation have a more important effect on selectivity. For example, Y zeolite exchanged with RE or with $H^{\oplus}$ ions produces gasolines of similar octane but La-exchanged forms produce more olefins and aromatics [200−202]. Studies by Corma et al. [203] on the effect of the cation in Y zeolite selectivity for isomerization, disproportionation, cracking, and the paraffin-to-olefin ratio in alkane cracking show that the main factors responsible for selectivity differences are the acid strength distribution, the Brønsted/Lewis ratio, and geometric factors connected with the zeolite structure [204]. Such factors are controllable by activation procedures, the Si/Al ratio of the material, and the choice of zeolite used.

When heavier feeds are to be cracked, not only are the selectivity considerations discussed previously of importance, but the additional problem of dealing with harmful metal poisons has to be confronted. The metals are present in heavy feeds in a number of organic compounds and are readily deposited on the catalyst by the cracking of these molecules. The effect of Ni, V, and Fe, which are the chief metallic components of gas oils, is to increase the yield of coke and dry gas while reducing gasoline yield. The unwanted activity of these metals is in the order of Ni > V > Fe and was bothersome even before the development of zeolite catalysts or the advent of heavy feed cracking [205−210]. To the delight of all concerned, zeolite catalysts have a better resistance to metals poisoning than do the old amorphous catalysts [207−214]. The dependence of the undesirable reactions is nonlinear with metal content of the catalyst [212] but correlates well with the half-power of the metal content [215]. Furthermore, there is no synergistic effect between the various metals. This seems to be due to the fact that various metals act in different ways. It seems that Ni does not poison cracking activity but rather catalyzes nonselective cracking, producing hydrogen, light ends, and coke. Vanadium becomes important as a producer of these products only at high (15,000 to 20,000 ppm) concentrations [215]. However, vanadium tends to migrate to the zeolite under the hydrothermal conditions of the regenerator and destroys the crystal structure, resulting in significant activity loss. The activity of the metal deposits decays with catalyst usage, perhaps due to sintering of the metal surface [208].

The relative resistance of zeolite catalysts to metals poisoning may be due to the operating conditions under which such catalysts are used rather than any inherent ability to tolerate such poisons. Zeolite catalysts are generally used at shorter contact times than were the amorphous catalysts. It has been proposed that this results in lower levels of metal sulfides being formed in the reactor and that it is the sulfides that cause the undesirable side reactions [208,216,217].

In any case the metals have undesirable effects on catalytic cracking, and several solutions to the problem have been proposed:

1. Demetallization of the feed

2. Compound formation with a special catalyst component

3. Poisoning the metal activity on the catalyst

In practice it is the last suggestion which has been most fruitful. "Passivating agents" which consist of compounds of antimony are added to circulating catalyst to poison metal activity [183–188, 218–221]. Bismuth [187] and tin [189] have also been successful in this role. The additives are introduced periodically or continuously to the feedstock and produce dramatic, though nonpermanent results. Alternative methods of metals removal such as the addition of chlorine-containing compounds in order to form volatile metal chlorides have foundered on corrosion and other technical difficulties [222].

3.7   GENERAL REMARKS ON CRACKING CATALYST BEHAVIOR

Today it is the silica-alumina catalysts which are almost universally used for catalytic cracking. With the advent of X and Y zeolites in suitably exchanged forms, most catalytic cracking processes have converted to various zeolite-containing amorphous silica-alumina catalysts for the cracking of hydrocarbons. The reason for this dramatic conversion of cracking processes to the use of zeolites is their superior activity and their selectivity for high-octane-gasoline-range products. The question that arises, is, therefore: What is the reason for this great difference between the activity and selectivity of amorphous silica alumina and that of zeolitic silica aluminas? One possible explanation could be that the acid centers on zeolites are of a different nature than those

on amorphous silica aluminas and that the mechanism of reaction
on these catalysts is somehow different from that on amorphous
materials.

This concept was analyzed by Eastwood et al. [197], who in
common with a number of other investigators found that the
activation energy of cracking for various paraffins is almost the
same on amorphous silica aluminas as it is on zeolites. This is
a strong indication that the active centers on both materials are
identical and that the mechanism of reaction is identical. In view
of this, perhaps it is differences in the number of active sites
that cause zeolite REX to crack normal hexadecane 17 times faster
than does amorphous silica alumina [223], or zeolite REY [224] to
be 10 times more active than amorphous silica alumina for the
cracking of gas oil, or the isomerization of xylenes to be some
40 times more rapid on LaHY [225].

When the concentration of acid sites is evaluated by n-butyl-
amine titration, one finds that the number of acid centers on the
HY zeolite is only some five times greater than that on amorphous
silica alumina [226]. Similar results on other zeolites show that
the difference observed in this way is not large enough to explain
the observed differences in activity.

In all fairness it should be pointed out that there are technical problems associated with butylamine titration, which measures
the total number of active sites on the zeolite, regardless of
whether they are Brønsted or Lewis sites and regardless of their
acid strength. At the same time it is well known that both Lewis
and Brønsted sites are present on silica aluminas and that the
distribution of acid strengths of both types of acids is quite
broad [3]. Furthermore, it is now widely recognized that not all
these acid sites take part in each reaction that is catalyzed by a
given catalyst. Indeed, it has been shown that the ratio of activities between zeolites and amorphous catalysts is dependent on
the test reaction used, such as normal hexadecane cracking, gas-
oil cracking, xylene isomerization, cumene cracking, and so on.
There is no doubt, however, that the differences in the rates of
the various reactions on various catalysts cannot be explained
solely by differences in total concentrations of active sites.

It has been suggested that the concentration of reactant in the
vicinity of the active sites differs between silica alumina and zeolites because the adsorption capacity of zeolites is much greater
than that of silica alumina. At the temperatures of catalytic
cracking, the concentration of hydrocarbons in the pores of a
zeolite is calculated to be some 50 times greater than that in
silica-alumina pores [227]. This, taken together with the site
concentration differences, may be enough to explain relative

activities, although obviously not in a quantitative way. However, difficulties are noted when one considers the cracking
activity of zeolites compared to silica alumina in the case of very
large molecules, ones that would have difficulty penetrating the
pores of a zeolite. In such cases zeolite catalysts are found to be
more active than silica alumina; internal concentrations and the
availability of sites which are postulated for zeolite catalysts
would appear to have no bearing on the matter [228]. A similar
objection can be raised to the electrostatic field hypothesis, which
suggests that the presence of electrostatic fields in the pores of
zeolites weakens bonds and enhances cracking rates in this manner [144].

More important than the differences in activity are the observed
differences in selectivity. Comparison of the distributions of the
products in the cracking of *n*-hexadecane can be found in the
work of Nace [223], whose results show that zeolites have much
greater selectivity for products of intermediate molecular weight;
that is, products which fall in the gasoline range. It is this
property of zeolites which has made them so valuable in commercial refining operations. Nace explains this phenomenon by considering the relative K_H/K_B values of zeolites and amorphous
catalysts (i.e., the ratio of the rate of hydrogen transfer to the
rate of cracking). He states that this ratio is greater for zeolites than it is for silica alumina, and that this results in the
stabilization of primary products on zeolite catalysts.

In general, the picture Nace proposes is therefore as follows.
Silica aluminas, which are poor hydrogen-transfer catalysts, lead
to overcracking of primary products to secondary products of
smaller molecular weight and to the formation of carbon; zeolites,
by facilitating hydrogen transfer, saturate product olefins and
arrest the reaction after the primary cracking step, resulting in
more primary products and less carbon production. His explanation for the higher hydrogen-transfer reaction velocity on zeolites
is based on the regular structure of zeolites, which tends to concentrate active sites in certain geometric configurations in close
proximity to one another, with the result that bimolecular reactions of the type involved in hydrogen transfer are enhanced by
the geometry of the catalyst lattice [229].

Another reason advanced for the differences in selectivity is
the difference in the distribution of acid site strengths. For instance, in silica alumina the majority of acid sites is found in the
region $H_0 < -8$, with some reported values being as high as $H_0 \sim$
-12, whereas in zeolites the acid strength distribution lies mainly
between -8 and -4 [167]. Moscou and Mone [72] have studied
the influence of acid strength on the selectivity of gas-oil cracking.

These authors report that the strongest acid sites are responsible for coke formation and the formation of light gases, whereas the acid sites of intermediate strength are responsible for
reactions leading to hydrogen transfer and desirable products.
This is supported by the fact that the dealumination of zeolite
has been shown to result in a reduction in the number of weak
acid sites and the formation of intermediate acid sites [231,232].
Topchieva et al. [233], for example, report that this procedure
changes selectivity in the direction of desirable products, sup-
porting the hypothesis that sites of moderate strength lead to the
formation of desirable products. Furthermore, this evidence confirms that there is a relationship between specific reactions and
acid sites of specific strength.

The remaining difference between zeolites and silica aluminas
is the crystalline structure of the zeolites, which gives zeolite
pores a unique configuration and a regularity not encountered in
amorphous materials. For instance, in the case of the zeolites X
and Y, the pore diameter is uniform and approximately 9 Å. This
regular pore structure is a factor that needs to be examined as to
its influence on activity and selectivity when zeolites are compared to silica alumina.

Many authors have commented on the influence of pore size
and shape on catalytic reactions, and in general many observed
effects have been ascribed to "shape-selective" catalysis. Under
this one all-encompassing appellation, one can readily distinguish
two different phenomena. One is the simple sieving effect due to
the pore size of the zeolite involved. The other is the spatial
distribution of active acid sites inside the pores and cavities of
the zeolites.

The sieving effect can be clearly shown by using a 5-Å zeolite
as a cracking catalyst [234]. This zeolite is able to crack normal
paraffins but cannot crack methylalkanes to any significant extent
due to the fact that methylalkanes are too large to enter the 5-Å
pores. In zeolites with larger pore sizes, such as X and Y, as
well as in silica alumina, both types of paraffins are readily
cracked. Such size selectivity has also been reported in the case
of H mordenite containing various large ions in the channels [235].
For example, barium H mordenite can crack normal and monomethyl
paraffins selectively out of a mixture of polymethyl paraffins.
Thus, for a molecule that is too large to enter the pores of a
zeolite, the only sites that are available are those situated on the
external surface of the zeolite crystals. Such sites represent a
very small fraction of the total acidity of the catalyst.

On closer examination of this phenomenon it becomes obvious
that there are intermediate situations between no restriction to

entry and total restriction as described above. In such intermediate situations, counterdiffusion is going to play an important role in the selectivity of the catalyst, due to the restrictions that are forced upon secondary reactions by the channel size and diffusion of products. For example, it has been shown that molecules such as n-hexane show the same activation energy when cracked on faujasite-type zeolites or on silica aluminas [148,236]. However, the activation energy is about one-half of that reported for the cases above when a small pore zeolite is used as a cracking catalyst [234].

In the case of gas-oil cracking, Thomas and Barmby [237] have suggested that because of the size of the reacting molecules, the initial reaction should occur on the surface of the zeolite crystals or on the surrounding silica-alumina matrix of a commercial catalyst. It is only the smaller primary products that can diffuse and penetrate into the crystalline structure, where hydrogen-transfer reactions take place. This idea points in the direction where we should seek a fuller explanation of the influence of molecular sieving.

The predominance of molecular species in the gasoline range is probably due to the fact that the carbon which is most likely to form a carbenium ion on a long chain snaking past an acid site is an interior carbon starting somewhere around the fourth or fifth one along the chain, because these carbons require lower ionization energies than do those carbons nearer the end of the chain. Carbons farther along the molecular chain require essentially the same energy for carbenium ion formation as does the fourth or fifth carbon. If a carbenium ion arises on the fourth or fifth carbon, the resultant olefin is five or six carbons long, which is exactly the length encountered in the lower range of gasoline molecules. The subsequent entrapment of the short-chain olefin in the interior of the zeolite is clearly conducive to a multitude of encounters with active sites and to the rearrangement of this olefin to branched or aromatic species, which are so desirable for high octane.

The shape selectivity that may be associated with the spatial distribution of acid sites in the interior of the zeolites can also be important in the formation of the observed products. It is now well known that ZSM-5 zeolite is able to produce a high proportion of aromatic compounds from methanol [238—242] or olefins [243]. Since the reactions necessary for this process are obviously not monomolecular, it is easy to see that the distribution of active sites in the pores of ZSM-5 must play a significant role in the formation of aromatics. Furthermore, contrary to the normal rules of alkylation, benzene alkylation is faster in the presence of ZSM-5 than is the alkylation of toluene [244,245]. This suggests that another effect is involved in shape selectivity. The first effect was due to the distances separating the sites and their spatial distribution; the other is a limitation on the molecular size of products that can be formed inside the channels of the zeolite. A good example of the effect of space limitations on reaction products is the failure of methylethylbenzene to disproportionate to trialkylbenzene on H mordenite. This appears to be due to the fact that the diphenyl intermediate cannot be formed inside the mordenite channel [246].

Wang et al. [247] report that the cracking of *n*-hexane on HZSM-5 produces greater yields of ethylene and of aromatics than does cracking on HY, H mordenite, or LaY zeolites. Unfortunately, the level of conversion at which the selectivity pattern was observed was too high for firm conclusions to be drawn. When the selectivity patterns of all these catalysts are compared at lower levels of a hexane conversion, no differences are observed in the C_5+ fraction on any of the catalysts noted above [247]. This illustrates one of the problems often encountered in catalysis research: conclusions are drawn and important suggestions made on the basis of experiments at high levels of conversions and at long times on stream, when the undefined properties of the decayed catalyst and the confusion caused by secondary reactions can grossly distort the more meaningful picture that would be observed at low conversions on a fresh catalyst.

To summarize the discussion, one can compare the activity of catalysts by looking at the reaction rate r as a function of several variables:

$$r = \Sigma \; f(k_0[acid \; sites],[S])$$

where k_0 is a rate constant whose magnitude is influenced by the nature of the active sites and by such ancillary effects as the deformations caused by steric effects. The second term refers to the concentration of acid sites and is controlled by parameters such as the ratio of silica to alumina, the level of exchange, the type of cation exchanged, and the nature of pretreatment of the catalyst—in other words, by all factors that affect the number and strength distribution of acid sites. The third term accounts for the concentration of reactant in the neighborhood of the active site. This term accounts for the size of the pores, adsorption effects, and of course concentration effects due to the bulk concentration of the reactant in the gas phase. A summation must then be made over all the types of sites and all molecular species present.

There is a growing tendency to compare the different factors affecting cracking catalyst activity and selectivity point by point. This is the result of an increasing awareness among researchers of the fact that no one simple effect is responsible for all the observed phenomena and for the differences between silica-alumina and zeolite catalysts [248−251]. It now appears that a given feed molecule may well react on a variety of sites, each leading to different products. Thus by changing the distribution of acid strengths on a catalyst and the proportion of Brønsted to Lewis sites, one can vary the product selectivity from a given feed-stock over a wide range. Factors such as the level of exchange, ratio of silica to alumina, pore size, structure, catalyst pretreatment, and so on, also contribute by means of effects discussed above and lead to product distributions which vary significantly from catalyst to catalyst without changing the kinetics of any individual reaction. Once we understand the effects of the various factors, we may be in a position to achieve the desired conversions by controlling variables during the synthesis, activation, and exchange of the catalyst.

To begin to unravel these various effects, we turn to the influence of catalyst decay on the observable results obtained in catalytic cracking studies. This has been the single most important roadblock to progress in understanding the fundamental behavior of catalytic cracking. By learning how to deal with catalyst decay in a quantitative way, we can begin to answer many of the questions that remain.

REFERENCES

1. A. N. Sachaven, *Conversion of Petroleum*, New York, Reinhold, 1948

2. M. Ocelli, *J. Catal.*, *90*: 256 (1984)

3. P. Courty and C. Marcilly, *Preparation of Catalysts III*, p. 485, Amsterdam, Elsevier, 1983

4. L. L. B. Ryland, M. W. Tamele, and J. N. Wilson, *Catalysis* (P. H. Emmett, Ed.), Vol. VII, p. 1, New York, Reinhold, 1960

5. J. J. Fripiat, *12th Natl. Conf. Clays Clay Minerals*, p. 327, 1963

6. C. de Kimpl, M. C. Gastuche, and G. W. Brindley, *Am. Mineral.*, *46*: 1370 (1961)

7. C. J. Plank and L. C. Drake, *J. Colloid Sci.*, *2*: 399 (1947)

8. M. R. S. Manton and J. C. Davidtz, *J. Catal.*, *60*: 156 (1979)

9. J. S. Magee, Jr. and R. P. Daugherty, Jr., *U.S. Patent* *3,912,619*, 1975

10 D. E. W. Vaughan, P. K. Maher, and E. W. Albers, *U.S. Patent 3,838,037*, 1974

11. T. O. Mitchell and D. D. Withehorst, *U.S. Patent 4,003,825*, 1977

12. P. G. Rouxhet and R. E. Semples, *J. Chem. Soc. Faraday, Trans. I*, *70*: 2021 (1974)

13. R. E. Semples and P. G. Rouxhet, *J. Colloid Interface Sci.*, *55*: 263 (1976)

14. P. O. Scokart, F. D. Declerck, R. E. Semples, and P. G. Rouxhet, *J. Chem. Soc., Faraday Trans. 1*, *73*: 359 (1977)

15. J. F. Damon, B. Delmon, and J.-M. Bonnier, *J. Chem. Soc., Faraday Trans. 1*, *73*: 372 (1977)

16. C. Defosse, P. Canesgon, P. G. Rouxhet, and B. Delmon, *J. Catal.*, *51*: 269 (1978)

17. J. B. Peri, *J. Catal.*, *41*: 227 (1976)

18. C. J. Plank, *Adv. Chem. Ser.* (Heterogeneous Catalysis), *222*: 253 (1983)

19. W. M. Meier and D. H. Olson, *Adv. Chem. Ser.* (Mol. Siev. Zeolites—I), *101*: 155 (1971); W. M. Meier and H. J. Mock, *J. Solid State Chem.*, *27*: 349 (1979)

20. D. E. W. Vaughan, *Natural Zeolites: Occurrence Properties and Use* (L. B. Sand and F. A. Mumpton, Eds.), London, Pergamon, 1978

21. W. M. Meier and D. H. Olson, *Atlas of Zeolite Structure Type*, Int. Zeolite Assoc. (1978)

22. R. M. Barrer, *Advanced Studies in Zeolites*, NATO Inst., Portugal, 1983

23. E. M. Flanigen, *Proc. Int. Conf. Mol. Sieve Zeolites,* *Naples,* p. 760, 1980

24. J. V. Smith, *Adv. Chem. Ser.* (Mol. Sieve Zeolites—I), *101:* 171 (1971)

25. W. J. Mortier, *Compilation of Extra Framework Sites in Zeolites,* Guilford, England, Butterworth Scientific Ltd., 1982

26. J. V. Smith, *Zeolite Chemistry and Catalysis* (J. Rabo, Ed.), Washington, D.C., Am. Chem. Soc., 1976

27. H. S. Sherry, *J. Phys. Chem.,* *70:* 1158 (1966)

28. H. S. Sherry, *Ion Exchange,* Vol. 2 (J. A. Marinsky, Ed.), p. 89, New York, Marcel Dekker, 1969

29. J. A. Rabo, P. E. Pickert, and J. E. Boyle, *U.S. Patent 3,130,006,* 1964

30. P. K. Maher and C. V. McDaniel, *U.S. Patent 3,402,996,* 1968

31. R. M. Milton, *U.S. Patent 2,882,244,* 1959

32. D. W. Breck, *U.S. Patent 3,130,007,* 1964

33. J. C. Pitman and L. J. Raid, *U.S. Patent 3,473,589,* 1969

34. D. W. Breck, *Zeolite Molecular Sieves,* New York, Wiley, 1974

35. D. W. Breck, E. M. Flanigen, and R. M. Milton, *Abstr. 137th Meeting Am. Chem. Soc.,* Apr. 1960

36. H. Robson, *Chem. Tech.,* *8:* 176 (1978)

37. L. D. Rollmann, *Adv. Chem. Ser.* (Inorg. Cmpd. Unusual Prop.—2), *173:* 387 (1979)

38. E. M. Flanigen, *Pure Appl. Chem.,* *52:* 2191 (1980)

39. L. B. Sand, *Pure Appl. Chem.,* *52:* 2105 (1980)

40. M. Mengel, *Chem. Tech.* (Heidelberg), *10:* 1135 (1981)

41. R. M. Barrer, *Zeolites,* *1:* 130 (1981)

42. L. D. Rollmann, *NATO ASI Ser.,* Ser. E, 1984, 80 (Zeolites: Sci. Technol.), Portugal, p. 109, 1983

43. D. W. Breck and E. M. Flanigen, *Int. Conf. Mol. Sieve Zeolites, London,* p. 47, 1967

44. C. Baerlocher and W. M. Meier, *Helv. Chim. Acta,* *52:* 1853 (1969)

45. R. J. Argauer and G. R. Landolt, *U.S. Patent 3,702,886*, 1972

46. R. H. Daniels, G. T. Kerr, and L. D. Rollmann, *J. Am. Chem. Soc.*, *100*: 3097 (1978)

47. E. E. Jenkins, *U.S. Patent 3,578,398*, 1971

48. E. Freund, *J. Cryst. Growth*, *34*: 11 (1976)

49. H. Kacirek and H. Lechert, *J. Phys. Chem.*, *79*: 1589 (1975)

50. J.-L. Guth, P. Caullet, and R. Wey, *Bull. Soc. Fr. Mineral. Crystallogr.*, *99*: 21 (1976)

51. T. J. Weeks and D. E. Passoja, *Clays Clay Miner.*, *25*: 211 (1977)

52. J. F. Tempere, D. Delafosse, and J. P. Contour, *Molecular Sieves II* (J. R. Katzer, Ed.), p. 76 (ACS Symp. Ser., No. 40), 1977

53. J. Klinowski, J. M. Thomas, C. A. Fyfe, and G. C. Gobbi, *Nature*, *296*: 533 (1982)

54. C. A. Fyfe, G. C. Gobbi, J. S. Hartman, J. Klinowski, and J. M. Thomas, *J. Phys. Chem.*, *86*: 1247 (1982)

55. C. A. Fyfe, G. C. Gobbi, J. S. Hartman, R. E. Lekinski, J. H. O'Brien, E. R. Beange, and M. A. R. Smith, *J. Mag. Reson.*, *47*: 168 (1982)

56. C. R. Adams and H. H. Voge, *J. Phys. Chem.*, *61*: 722 (1957)

57. W. G. Schlaffer, C. Z. Morgan, and J. N. Wilson, *J. Phys. Chem.*, *61*: 714 (1957)

58. M. S. Goldstain, *Experimental Methods in Catalytic Research* (R. B. Anderson, Ed.), p. 361, New York, Academic Press, 1968

59. K. M. Elliott and S. C. Eastwood, *Proc. Div. Refin.*, *Am. Pet. Inst.*, *42*: 272 (1962)

60. C. J. Plank, E. J. Rosinski, and W. P. Hawthorne, *Ind. Eng. Chem.*, *Prod. Res. Dev.*, *3*: 165 (1964)

61. C. J. Plank and E. J. Rosinski, *Chem. Eng. Prog. Symp. Ser.*, *63*(73): 26 (1967)

62. C. V. McDaniel and P. K. Maher, *Zeolite Chemistry and Catalysis* (J. A. Rabo, Ed.), p. 171 (ACS Monogr. 285), 1976

63. R. M. Barrer and D. A. Langley, *J. Chem. Soc.*, 3804 (1958)

64. R. W. Baker, P. K. Maher, and J. J. Blazek, *Hydrocarbon Process.*, 47(2): 125 (1968)

65. P. B. Venuto and E. T. Habib, *Catal. Rev. Sci. Eng.*, *18*: 1 (1978)

66. P. K. Maher and C. V. McDaniel, *U.S. Patent 3,293,192*, 1966

67. C. V. McDaniel and P. K. Maher, *U.S. Patent 3,449,070*, 1969

68. G. T. Kerr, *J. Catal.*, *15*: 200 (1969)

69. P. Jacobs and J. B. Uytterhoeven, *J. Catal.*, *22*: 193 (1971)

70. G. T. Kerr, *J. Phys. Chem.*, *72*: 2594 (1968)

71. G. T. Kerr, *J. Phys. Chem.*, *73*: 2780 (1969)

72. L. Moscou and R. Mone, *J. Catal.*, *30*: 417 (1973)

73. L. Forni, *Catal. Rev. Sci. Eng.*, *8*: 65 (1973)

74. H. A. Benesi and B. H. C. Winguist, *Adv. Catal.*, *27*: 97 (1978)

75. P. A. Jacobs, Ed., *Carboniogenic Activity of Zeolites*, Amsterdam, Elsevier, 1977

76. T. R. Hughes and H. M. White, *J. Phys. Chem.*, *71*: 2192 (1967)

77. H. A. Benesi, *J. Catal.*, *28*: 176 (1973)

78. P. A. Jacobs and C. F. Heylen, *J. Catal.*, *10*, 343 (1974)

79. A. V. Kiselev, D. G. Kitiashvili, and V. I. Lygin, *Kinet. Katal.*, *12*: 1075 (1971) [*12*, 959 (1971) in English edition]

80. J. Datka, *J. Chem. Soc., Faraday Trans*, 1, 77: 511

81. Y. Kageyama, T. Yotsuyanagi, and K. Aomura, *J. Catal.*, *36*: 1 (1975)

82. B. D. Flockhart, M. C. Megarry, and R. C. Pink, *Adv. Chem. Ser.* (Mol. Sieves, Int. Conf., 3rd), *121*: 509 (1973)

83. S. P. Zhdanov and E. I. Kotov, *Adv. Chem. Ser.* (Mol. Sieves, Int. Conf., 3rd), *121*: 240 (1973)

84. A. Abou-Kais, J. Vedrine, J. Massardier, and G. Dalmai-Imelik, *J. Catal.*, *34*: 317 (1974)

85. D. Ballivet, J. C. Vedrine, and D. Barthomeuf, *C. R. Acad. Sci.*, Ser. C, *283*: 429 (1976)

86. M. M. Mestdagh, W. E. E. Stone, and J. J. Fripiat, *J. Phys. Chem.*, *76*: 1220 (1972); *J. Catal.*, *38*: 358 (1975); *J. Chem. Soc.*, *Faraday Trans. 1*, *72*: 154 (1976)

87. D. Freude, W. Oehme, W. Schmiedel, and B. Staudte, *J. Catal.*, *32*: 137 (1974)

88. A. Corma, V. Fornes, and C. Rodenas, *J. Catal.*, *88*: 374 (1984)

89. A. K. Gosh and G. Curthoys, *J. Chem. Soc.*, *Faraday Trans. 1*, *80*: 99 (1984)

90. C. Kemball, H. F. Leach, B. Skundric, and K. C. Taylor, *J. Catal.*, *27*: 416 (1972)

91. T. Shiba, M. Sato, H. Hattori, and K. Yoshida, *Shokubai*, *6*(2): 80 (1964)

92. E. P. Parry, *J. Catal.*, *2*: 371 (1961)

93. M. R. Basila, T. R. Kantner, and K. H. Ree, *J. Phys. Chem.*, *68*: 3197 (1964)

94. M. R. Basila and T. R. Kantner, *J. Phys. Chem.*, *70*: 1681 (1966)

95. D. Ballivet, D. Barthomeuf, and P. Pichat, *J. Chem. Soc.*, *Faraday Trans. 1*, *68*: 1712 (1972)

96. J. A. Schwarz, B. G. Russell, and H. F. Harnsberger, *J. Catal.*, *54*: 303 (1978)

97. H. P. Leftin and W. K. Hall, *Proc. 2nd Int. Congr. Catal.*, *Paris*, *1*: 1353 (1960)

98. B. D. Flackhart and R. C. Pink, *J. Catal.*, *4*: 90 (1965)

99. H. R. Gerberich and W. K. Hall, *J. Catal.*, *3*: 113 (1964)

100. J. B. Peri, *J. Phys. Chem.*, *70*: 3168 (1966)

101. J. L. Garcia Fierro, A. Corma, F. Tomas, and R. Montainana, *Z. Phys. Chem.*, *127*: 87 (1982)

102. R. W. Maatman, D. L. Leenstra, A. Leenstra, R. L. Blankespoor, and D. N. Rubingh, *J. Catal.*, 7: 1 (1967)

103. W. B. Horton and R. W. Maatman, *J. Catal.*, 3: 113 (1964)

104. K. H. Bourne, F. R. Cannings, and R. C. Pitkethly, *J. Phys. Chem.*, 75: 220 (1971)

105. M. Marczewski and B. W. Wojciechowski, *Can. J. Chem. Eng.*, 60: 617 (1982)

106. J. Take, H. Kawai, and Y. Toneda, *J. Catal.*, 36: 356 (1975)

107. S. H. C. Liang and Ian D. Gay, *J. Catal.*, 66: 294 (1980)

108. J. Take, T. Tsuruya, T. Sato, and Y. Yoneda, *Bull. Chem. Soc. Jpn.*, 45: 3409 (1972)

109. J. A. Rabo, P. E. Pickert, D. N. Stamires, and J. E. Boyle, *Proc. 2nd Int. Congr. Catal.*, Paris, 2: 2055 (1960)

110. J. B. Uytterhoeven, L. G. Christner, and W. K. Hall, *J. Phys. Chem.*, 69: 2117 (1965)

111. R. L. Stevenson, *J. Catal.*, 21: 113 (1971)

112. G. T. Kerr, *J. Catal.*, 77: 307 (1982)

113. A. Corma, A. Lopez Agudo, and V. Fornes, *J. Chem. Soc., Chem. Commun.*, 942 (1983)

114. A. P. Bolton and M. A. Lanewala, *J. Catal.*, 18: 154 (1970)

115. J. W. Ward and R. C. Hansford, *J. Catal.*, 13: 364 (1964)

116. J. Turkevich, F. Nozaki, and D. Stamires, *Proc. 3rd Int. Congr. Catal.*, Amsterdam, 1: 586 (1964)

117. R. Beaumont, D. Barthomeuf, and Y. Trambouze, *Adv. Chem. Ser.* (Mol. Siev Zeolites—II), 102: 327 (1971)

118. A. Corma, H. Farag, and B. W. Wojciechowski, *Int. J. Chem. Kinet.*, 13: 883 (1981)

119. A. Lopez Agudo, A. Asensio, and A. Corma, *Can. J. Chem. Eng.*, 60: 50 (1982)

120. J. W. Ward, *J. Catal.*, 18: 348 (1970)

121. P. A. Jacobs and J. B. Uytterhoeven, *J. Chem. Soc., Faraday Trans. 1*, 69: 373 (1973)

122. J. Scherzer and J. L. Bass, *J. Catal.*, 28: 101 (1973)

123. A. Corma and V. Fornes, *Zeolites*, *3*: 197 (1983)

124. P. E. Eberly, Jr., *J. Phys. Chem.*, *72*: 1042 (1968)

125. R. Beaumont and D. Barthomeuf, *J. Catal.*, *26*: 218 (1972)

126. D. Barthomeuf, *Acta Phys. Chem.*, *24*: 71 (1978)

127. P. A. Jacobs, *Catal. Rev. Sci. Eng.*, *24*: 415 (1982)

128. L. A. Pine, P. J. Maher, and W. A. Wachter, *J. Catal.*, *85*: 466 (1984)

129. S. Beran and J. Dubsky, *J. Phys. Chem.*, *83*: 2538 (1979)

130. K. G. Ione, V. G. Stepanov, G. V. Echevskii, A. A. Shubin, and E. A. Paukshtis, *Zeolites*, 4: 114 (1984)

131. V. B. Kazanski, *Stud. Surf. Sci. Catal.* (Struct. React. Mod. Zeolites), *18*: 61 (1984)

132. H. Lohse, E. Aldsdorf, and H. Stach, *Z. Anorg. Allg. Chem.*, *447*: 64 (1978)

133. G. H. Kul and H. S. Sherry, *Proc. 5th Int. Conf. Zeolites, Naples* (L. V. C. Rees, Ed.), p. 813, London, Mayden, 1980

134. H. K. Beyer and I. Belenykaja, *Stud. Surf. Sci. Catal.* (Catal. Zeolites), Amsterdam, Elsevier, *5*: 203, 1980

135. J. Klinowski, J. M. Thomas, M. Audier, and S. Vasuderan, *J. Chem. Soc., Chem. Commun.*, 570 (1981)

136. D. Barthomeuf and R. Beaumont, *J. Catal.*, *30*: 288 (1973)

137. D. Barthomeuf and R. Beaumont, *J. Catal.*, *34*: 327 (1974)

138. E. Dempsey, *J. Catal.*, *33*: 497 (1974)

139. D. H. Olson, W. O. Haag, and R. M. Lago, *J. Catal.*, *61*: 390 (1980)

140. W. O. Haag, R. M. Lago, and P. B. Weisz, *Nature*, *309*: 589 (1985)

141. B. Beagley, J. Dwyer, F. R. Fitch, R. Mann, and J. Walters, *J. Phys. Chem.*, *88*: 1744 (1984)

142. K. V. Topchieva and H. S. Thuoang, *Kinet. Katal.*, *11*: 490 (1970); *12*: 1203 (1971)

143. R. M. Barrer, *Proc. 5th Int. Conf. Zeolites, Naples* (L. V. C. Ress, Ed.), p. 273, London, Hayden, 1980

144. J. A. Rabo, C. L. Angell, P. H. Kasai, and V. Schomaker, *Discuss. Faraday Soc.*, *41*: 328 (1966)

145. J. W. Ward, *J. Catal.*, *10*: 34 (1968)

146. A. K. Cheetham, M. M. Eddy, and J. M. Thomas, *J. Chem. Soc.*, *Chem. Commun.*, *20*: 1337 (1984)

147. S. Hocevar and B. Drzaj, *J. Catal.*, *73*: 205 (1982)

148. J. N. Maile, N. Y. Chen, and P. B. Weisz, *J. Catal.*, *6*: 278 (1966)

149. A. A. Spozhakina, I. F. Moskovskaya, and K. V. Topchieva, *Kinet. Katal.*, *8*: 614 (1967)

150. K. V. Topchieva, B. V. Romanovskii, L. I. Piguzova, H. S. Thoang, and Y. W. Bizreh, *Proc. 4th Int. Congr. Catal.*, *Moscow*, p. 57, 1968

151. R. Mone and L. Moscou, *Adv. Chem. Ser.* (Mol. Sieves, Int. Conf., 3rd), *121*: 351 (1973)

152. D. Ballivet, P. Pichat, and D. Barthomeuf, *Adv. Chem. Ser.* (Mol. Sieves, Int. Conf., 3rd), *121*: 469 (1973)

153. D. Barthomeuf, *NATO ASI Ser.*, Ser. E, 1984, 80 (Zeolites: Sci. Technol.), Portugal, p. 317, 1983

154. J. Dwyer, F. R. Fitch, and E. E. Nkang, *J. Phys. Chem.*, *87*: 5402 (1983)

155. E. Dempsey, *J. Catal.*, *39*: 155 (1975)

156. R. J. Mikovsky and J. F. Marshall, *J. Catal.*, *44*: 170 (1976)

157. G. Engelhardt, D. Zeigan, E. Lippmaa, and M. Magi, *Z. Anorg. Allg. Chem.*, *468*: 35 (1980)

158. G. Engelhardt, U. Lohse, A. Samoson, M. Magi, M. Tarmak, and E. Lippmaa, *Zeolites*, *2*: 59 (1982)

159. W. J. Mortier and P. Geerlings, *J. Phys. Chem.*, *84*: 1982 (1980)

160. D. Barthomeuf, *J. Chem. Soc.*, *Chem. Commun.*, 743 (1977)

161. R. Beaumont, P. Pichat, D. Barthomeuf, and Y. Trambouze, *Proc. 5th Int. Congr. Catalysis, Amsterdam*, *1*: 343 (1972)

162. J. Datka, *J. Chem. Soc.*, *Faraday Trans. 1*, *77*: 2877 (1981)

163. E. M. Gladrow, R. W. Krebs, and C. N. Kimberlin, Jr., *Ind. Eng. Chem.*, *45*: 142 (1953)

164. K. M. Elliot and S. C. Eastwood, *Oil Gas J.*, *60*(23): 142 (1962)

165. S. C. Eastwood, R. D. Drew, and F. D. Hartzell, *Oil Gas J.*, *60*(44): 152 (1962)

166. C. E. Johning and H. Z. Martin, *Adv. Chem. Ser.* (Heterogeneous Catalysis), *222*: 273 (1983)

167. A. G. Oblad, *Oil Gas J.*, *70*(13): 84 (1972)

168. R. B. Secor, *U.S. Patent 3,446,727*, 1969

169. R. B. Secor and E. S. Peer, *U.S. Patent 2,935,463*, 1960

170. H. E. Robson, *U.S. Patent 4,098,676*, 1978

171. B. Drzaj, V. Pirnat-Smuc, and S. Hocevar, *Vestn. Slov. Kem. Drus.*, *27*(1): 5 (1980)

172. J. S. Magee and J. J. Blazek, ACS Monogr. 171, *Zeolite Chem. Catal.* (J. A. Rabo, Ed.), p. 615, 1976

173. W. L. Haden, Jr., and F. J. Dzierzanowski, *U.S. Patent 3,657,154*, 1972

174. W. L. Haden, Jr., and F. J. Dzierzanowski, *U.S. Patent 3,663,165*, 1972

175. M. G. Howden, *CSIR Rep.* (Series), CENG 365, 1981

176. L. Rheaume, R. E. Ritter, J. J. Blazek, and J. A. Montgomery, *Oil Gas J.*, *74*(20): 103 (1976)

177. A. W. Chester, A. B. Schwartz, W. A. Stover, and J. P. McWilliams, *Prepr., Div. Pet. Chem., Am. Chem. Soc.*, *24*: 624 (1979)

178. F. D. Hartzell and W. A. Chester, *Hydrocarbon Process.*, *58*(7): 137 (1979)

179. R. J. Bertolacini, G. M. Lehmann, and E. G. Wollaston, *U.S. Patent 3,835,031*, 1974; S. Jin and J. A. Jaecker, *U.S. Patent 4,472,267*, 1984; J. W. Bryne, *Nat. Pet. Ref. Assoc.*, AM-84-55, 1984

180. I. A. Vasalos, E. R. Strong, C. K. R. Hsieh, and G. J. D'Souza, *Oil Gas J.*, *75*(26): 141 (1977)

181. D. P. McArthur, H. D. Simpson, and K. Baron, *Oil Gas J.*, *79*(8): 55 (1981)

182. J. M. Maselli and A. W. Peters, *Catal. Rev. Sci. Eng.*, *26*: 525 (1984)

183. G. H. Dale and D. L. McKay, *Hydrocarbon Process.*, *56*(9): 97 (1977)

184. D. L. McKay, *U.S. Patents 4,025,458* and *4,031,002*, 1977

185. D. L. McKay and B. J. Bertus, *Prepr.*, *Div. Pet. Chem.*, *Am. Chem. Soc.*, *24*: 645 (1979)

186. M. B. Neuworth and E. C. Moroni, *Adv. Coal Util. Technol. Symp. Pap.*, Chicago, IGT, p. 345, 1979

187. T. C. Readal, J. D. McKinney, and R. A. Titmus, *U.S. Patent 3,977,963*, 1976

188. A. W. Chester, *Prepr.*, *Div. Pet. Chem.*, *Am. Chem. Soc.*, p. 505, 1981

189. J. D. McKinney and B. R. Mitchell, *U.S. Patent 4,040,945*, 1977

190. C. J. Plank and L. C. Drake, *J. Colloid Sci.*, *2*: 399 (1947)

191. K. D. Ashley and W. B. Innes, *Ind. Eng. Chem.*, *44*: 2857 (1952)

192. A. L. Hensley and J. E. Barney, *J. Phys. Chem.*, *62*: 1560 (1958)

193. R. M. Dobres, L. Rheaume, and F. G. Ciapetta, *Ind. Eng. Chem.*, *Prod. Res. Dev.*, *5*: 174 (1966)

194. C. P. Wilson, B. Carr, and F. G. Ciapetta, *U.S. Patent 3,395,103*, 1968

195. B. de Kroes, *Akzo Chem. Rep.*, 1982

196. J. S. Magee, *ACS Symp. Ser.* (Molecular Sieves—II; Int. Conf., 4th), p. 650, 1977

197. S. C. Eastwood, C. J. Plank, and P. B. Weisz, *Proc. 8th World Petr. Congr.*, *4*: 245 (1971)

198. L. L. Upson, *Katal. Rep.*, 1982

199. N. Y. Chen, T. O. Mitchel, D. H. Olson, and B. P. Perlrine, *Ind. Eng. Chem.*, *Prod. Res. Dev.*, *16*: 244 (1977)

200. J. S. Magee and R. E. Ritter, *Prepr.*, *Div. Pet Chem.*, *Am. Chem. Soc.*, *23*: 1057 (1978)

201. J. Sherzer and R. E. Ritter, *Ind. Eng. Chem., Prod. Res. Dev., 17:* 219 (1978)

202. W. J. Reagan, G. M. Wolterman, and S. M. Brown, *Prepr., Div. Pet. Chem. Am. Chem. Soc., 28:* 884 (1983)

203. A. Corma, J. B. Monton, and A. V. Orchilles, *Ind. Eng. Chem., Prod. Res. Dev., 23:* 404 (1984)

204. A. Corma, J. H. Planelles, and F. Tomas, *J. Catal., 94:* 425 (1985)

205. G. A. Mills, *Ind. Eng. Chem., 42:* 182 (1950)

206. B. J. Duffy and H. M. Hart, *Chem. Eng. Prog., 48:* 344 (1952)

207. E. C. Gossett, *Pet. Refiner, 39*(6): 177 (1960)

208. H. R. Grane, J. E. Conner, and G. P. Masologites, *Pet. Refiner, 40*(5): 168 (1961)

209. R. E. Donaldson, T. Rice, and J. R. Murphy, *Ind. Eng. Chem., 53:* 721 (1961)

210. R. G. Meisenheimer, *J. Catal., 1:* 356 (1962)

211. R. N. Cimbalo, R. L. Foster, and S. J. Wachtel, *Oil Gas J., 70*(20): 112 (1972)

212. G. P. Masologites and L. H. Beckberger, *Oil Gas J., 71*(47): 49 (1973)

213. E. T. Habib, Jr., H. Owen, P. W. Snyder, C. W. Streed, and P. V. Venuto, *Ind. Eng. Chem., Prod. Res. Dev., 16:* 291 (1977)

214. B. R. Mitchell, *Ind. Eng. Chem., Prod. Res. Dev., 19:* 209 (1980)

215. A. W. Chester, *Prepr., Div. Pet. Chem., Am. Chem. Soc., 26:* 505 (1981)

216. J. J. Blazek, *Oil Gas J., 71*(41): 65 (1973)

217. J. J. Blazek, *Catalagram, 42:* 3 (1973)

218. M. M. Johnson and D. C. Tabler, *U.S. Patent 3,711,422,* 1973

219. D. L. McKay, *U.S. Patent 4,031,002,* 1977

220. D. L. McKay and B. J. Bertus, *Prepr., Div. Pet. Chem., Am. Chem. Soc.,* p. 645, 1979

221. R. O. Day, M. M. Chauvin, and W. E. McEwen, *Phosphorus and Sulfur and Related Elements*, 8: 121 (1980)

222. R. M. Suggit and P. L. Paull, *U.S. Patent 4,013,546*, 1977

223. D. M. Nace, *Ind. Eng. Chem., Prod. Res. Dev.*, 8: 31 (1969)

224. D. M. Nace, *Ind. Eng. Chem., Prod. Res. Dev.*, 9: 203 (1970)

225. R. C. Hansford and J. W. Ward, *J. Catal.*, 13: 316 (1969)

226. H. Otouma, Y. Arai, and H. Ukihashi, *Bull. Chem. Soc., Jpn.*, 42: 2449 (1969)

227. P. B. Venuto, *Chem. Technol.*, 1: 215 (1971)

228. A. Corma and B. W. Wojciechowski, *Catal. Rev. Sci. Eng.*, 27: 29 (1985)

229. M. L. Poutsma, ACS Monogr. 171, *Zeolite Chem. Catal.* (J. A. Rabo, Ed.), p. 439, 1976

230. R. J. Mikovsky and J. F. Marshall, *J. Catal.*, 44: 170 (1976)

231. D. Barthomeuf, *J. Phys. Chem.*, 83: 249 (1979)

232. A. Auroux, P. C. Gravelle, J. C. Vedrine, and M. Rekas, *Proc. 5th Int. Conf. Zeolites, Naples* (L. V. Rees, Ed.), London, Heyden, p. 433, 1980

233. K. V. Topchieva, L. M. Vishnevskaya, and H. S. Thoang, *Dokl. Akad. Nauk SSSR*, 213: 1368 (1973)

234. P. B. Weisz, V. J. Frilette, R. W. Maatman, and E. B. Mower, *J. Catal.*, 1: 307 (1962)

235. T. Yashima, A. Yoshimura, and S. Namba, *Proc. 5th Int. Conf. Zeolites, Naples* (L. V. Rees, Ed.), London, Hayden, p. 705, 1980

236. N. Y. Chen, *Proc. 5th Int. Congr. Catal.*, Amsterdam, 2: 13433 (1972)

237. C. L. Thomas and D. S. Barmby, *J. Catal.*, 12: 341 (1968)

238. J. J. Wise and A. J. Silvestri, *Oil Gas J.*, 74(47): 140 (1976)

239. S. L. Meisel, J. P. McCullough, C. H. Lechthaler, and P. B. Weisz, *Chem. Technol.*, 6: 86 (1976)

240. C. D. Chang and A. J. Silvestri, *J. Catal.*, *47*: 249 (1977)

241. E. G. Derouane, J. B. Nagy, P. Dejaifve, J. H. C. Van Hoff, B. P. Spekman, J. C. Vedrine, and C. Naccache, *J. Catal.*, *53*: 40 (1978)

242. N. Y. Chang and W. J. Reagan, *J. Catal.*, *59*: 123 (1979)

243. E. N. Givens, C. J. Planck, and E. J. Rosinski, *U.S. Patent 3,960,978*, 1976

244. H. Heinemann, *Catal. Rev. Sci. Eng.*, *15*: 53 (1977)

245. W. E. Garwood and N. Y. Chen, *Prepr.*, *Div. Pet. Chem.*, *Am. Chem. Soc.*, *25*: 84 (1980)

246. P. B. Venuto, *Catalysis Org. Synt.* (Conf., 6th), p. 67, 1979

247. I. Wang, T.-J. Chen, K. J. Chao, and T.-C. Tsai, *J. Catal.*, *60*: 140 (1979)

248. J. A. Rabo and P. H. Kasai, *Prog. Solid State Chem.*, *9*: 1 (1975)

249. W. J. Mortier, *J. Catal.*, *55*: 138 (1978)

250. P. A. Jacobs, W. J. Mortier, and J. B. Uytterhoeven, *J. Inorg. Nucl. Chem.*, *40*: 1919 (1978)

251. D. Barthomeuf, *Stud. Surf. Sci. Catal.* (Catal. Zeolites) (B. Imelik et al., Eds.), *5*: 55 (1980)

4

Catalyst Decay and Selectivity Behavior

4.1   CAUSES AND KINETICS OF CATALYST DECAY

Most catalysts are subject to loss of activity during use [1,2]. In general, efforts to minimize this loss are successful, with the result that most commercial catalysts can be on stream for months or years before replacement becomes necessary. Not so in the case of cracking catalysts. The materials described in Chapter 3 all show a rapid decline of activity with time on stream. In fact, there seems to be an inverse relationship between activity and the rate of decay. This is largely why the early, low-activity clay catalysts were used under conditions where they spent minutes on stream while the newest high-activity zeolite catalysts spend seconds. The causes and kinetics of catalyst decay are therefore central to the improvement of cracking catalysts and to an understanding of cracking behavior.

4.1.1   Causes of Catalyst Decay

In general, catalyst deactivation can be divided into that due to chemical causes and that due to physical causes. The physical causes include sintering, occlusion, loss of surface area, and so on, and have been considered in Chapter 3. Catalysts that are commercially successful are normally quite stable physically under operating conditions. This is also true for cracking catalysts, whose physical stability would in principle allow operation for many weeks or months.

Chemical causes are more of a problem in general, and in particular, in catalytic cracking. To proceed further we can subdivide chemical deactivation as follows:

1. *Inhibition*, which is caused by the competitive adsorption of poisoning species present in the feed. The deactivation in this case is fully reversible by purging with clean feed, which leads to a constant but lower catalyst activity, and can be prevented by removing the poisoning species from the feed [3].

2. *Poisoning by impurities*, which is due to the nonreversible adsorption of impurities present in the feed. This leads to eventual loss of all activity and is dependent on the catalyst-to-reagent ratio. Poisoning by metals in catalytic cracking is an example of impurity poisoning, which, in the event, is hard to forestall by feed purification.

3. *Self-poisoning*, which is due to poisoning by the desired reaction. This is the main cause of catalyst decay in cracking
   and the reason for the elaborate and varied reactor-regen-
   erator designs which have evolved in the commercial applications of catalytic cracking.

The self-poisoning of cracking catalysts is accompanied by the
deposition of a carbonaceous material on the catalyst surface and
in the pore structure. This "coke" is an ill-defined product
whose composition can vary downward from an H/C ratio of 2, and
whose properties are dependent on the catalyst used, the feedstock, operating conditions, and stripping conditions. Various
authors have attempted to classify the total observed coke into
fractions that arise by various mechanisms. An example [4–7] is
the following conceptual breakdown of the total coke on catalyst:

1. *Coke of addition*: the dehydrogenated residue of the non-
   volatile fraction of the feed

2. *Contaminant coke*: the material produced, not by the catalyst
   itself, but by metal contaminants that had previously been
   deposited on the catalyst by the feed

3. *Catalyst-to-oil coke*: strippable material left in the pores of
   the catalyst by the feed

4. *Catalytic coke*: the coke produced by the main catalytic
   process, which is an inevitable part of the cracking reaction

These four types in themselves do not define coke any more
closely as to properties, but do give some indication of the variety
of materials present in coke, which, in any given situation, arise
from various sources. Table 4.1 gives one example of this distribution.

Clearly, substantial portions of the coke are from sources other
than the reaction itself. On the other hand, it seems intuitively
obvious that the coke loadings, which reach more than 10 wt % on
catalyst, must be responsible for, or associated with, the loss of
activity. Thus the questions that arise are: Is all coke equally
deactivating? Is the deactivation due to pore blocking, surface
blanketing, or diffusional resistance due to pore constriction
[8–11]?

Various attempts have been made to answer the question of how
coke is related to catalyst activity. The classical treatment of

Table 4.1   Coke Origins on an FCC
Catalyst

Type of coke	Percent of total coke
Conradson	5
Contaminant	30
Catalyst-to-oil	20
Catalytic	45

Source: Ref. 7.

Wheeler [12] considered the effects of pore constriction [13] on
both activity and selectivity. This elegant mathematical approach
did much to clarify thinking on the effects of diffusional limitations on catalytic reactions, but neither it nor its further exten-
sions were of great utility in describing cracking catalyst behavior in practice.

    Two other approaches have been used to describe decay. One
is based directly on measurement of coke on catalyst (COC) [14–
16] and the other is based on time on stream (TOS) [17,18].
These two have been more successful in dealing with real systems.
We begin with the more obvious approach, that using coke on
catalyst as a measure of catalyst activity.

4.1.2   Decay Models Based on Coke on Catalyst

Kinetic models of catalyst decay based on COC are attractive in
the first instance because they seem to relate loss of activity to
the amount of supposed deactivating agent. Unfortunately, the
variety of coke origins and the range of mechanisms by which
coke can cause deactivation make modeling by this approach very
cumbersome, both experimentally and mathematically.

    The major effort in developing the COC approach has been
that of Froment [19], who, in deriving his kinetic descriptions of
catalyst decay, has had to make significant simplifications of the
complex reality in the decay phenomena. The result is that this
treatment describes coke deposition and catalyst activity using
empirical concepts. Froment defines a_A as the "activity" of the
catalyst with respect to the main cracking reaction and a_C as the

activity with respect to coke-producing reactions. He then re-
lates these activities to the fraction of sites available for reaction:

$$a_A = \left(\frac{C_t - C_{cl}}{C_t}\right)^{n_A} \tag{4.1}$$

$$a_C = \left(\frac{C_t - C_{cl}}{C_t}\right)^{n_C} \tag{4.2}$$

where C_t is the total concentration of sites available for reaction
and C_{cl} is the concentration of poisoned sites. Equations (4.1)
and (4.2) express a model rather than any mechanistic reality and
in practice can be replaced by any suitable modeling function that
leads to a good fit of a given data set [14,20,21]. For example,
the functions

$$a_A = \exp(-\alpha_A C_C) \tag{4.3}$$

$$a_C = \exp(-\alpha_C C_C) \tag{4.4}$$

allow a connection to be made between the activities and C_C, the
coke on catalyst. Defining "activities" in this way leads to rate
expressions of the form

$$\frac{dC_X}{dt} = r^o_X \, a_A \tag{4.5}$$

$$\frac{dC_C}{dt} = r^o_C \, a_C \tag{4.6}$$

where

C_X = fraction of the feed converted

C_C = coke on catalyst

$r^o_{X,C}$ = rates of the respective reactions at zero coke level

t = time on stream

Introducing Equations (4.3) and (4.4) into (4.5) and (4.6) and integrating, one obtains

$$C_X = \frac{1}{\alpha_A} \ln(1 + \alpha_A r^o_X t) \qquad\qquad (4.7)$$

$$C_C = \frac{1}{\alpha_C} \ln(1 + \alpha_C r^o_C t) \qquad\qquad (4.8)$$

Equations (4.7) and (4.8) are clearly a long way from the initially hoped for mechanistic model of the effects of coke on catalyst activity. They can be fitted to experimental data to yield α_A, α_C, r^o_X, and r^o_C and hence can be used to describe coke profiles and activity behavior.

It is a pity that the complex nature of coke does not allow a more mechanistic interpretation of catalyst decay. Furthermore, coke measurements are time consuming and subject to experimental error, making the use of COC models of catalyst decay cumbersome to use in practice. This is compounded by the fact that coke on catalyst is a poor indicator of catalyst activity, as noted by Plank and Nace [22] in cumene cracking, by Heinemann [23] in the dehydrogenation of methylcyclopentane on $Cr_2O_3-Al_2O_3$, and by John et al. [24] in gas-oil cracking.

All in all, the advantages of COC models that were anticipated have not materialized in practice. Nevertheless, the methods of Froment are a valuable source of correlations for describing coke-on-catalyst profiles in reactors [16].

4.1.3   Decay Models Based on Time on Stream

The difficulties described in Section 4.1.2 with respect to achieving a mechanistic model of catalyst decay are such that many workers have resorted to the much more convenient and less mechanistic functional dependence of catalyst activity on time on stream. Perhaps the simplest such model involves the assumption that catalyst "activity" a is linearly dependent on the time on stream t [25]:

$$a = a_0 - At \tag{4.9}$$

whose differential form is zero order in activity.

$$-\frac{da}{dt} = A \tag{4.10}$$

More elaborate forms use first-order dependencies [26-28],

$$-\frac{da}{dt} = Aa \tag{4.11}$$

$$a = a_0 \exp(-At) \tag{4.12}$$

second-order dependencies [29,30],

$$-\frac{da}{dt} = Aa^2 \tag{4.13}$$

$$\frac{1}{a} = \frac{1}{a_0} + At \tag{4.14}$$

and other more involved forms, such as those used by Blanding [31] and Voorhies [32]:

$$-\frac{da}{dt} = BA^{1/2}a^{(B+1)/B} \tag{4.15}$$

$$\left(\frac{1}{a}\right)^{1/B} = const_1 + const_2(t) \tag{4.16}$$

All these forms were applied successfully in specific cases. In particular, the exponential form given in Equation (4.12) allowed Weekman [18] to develop models of catalytic cracking which were mathematically and esthetically pleasing in their simplicity [18].

All the decay functions above can be encompassed by the generalized forms developed in the time-on-stream theory [23], where it is postulated that the "activity" is in fact directly related to the concentration of active sites, C_S, on the catalyst. We begin by formulating an expression for the dependence of C_S:

$$- \frac{dC_S}{dt} = k'_{0d} + k'_{1d}C_S + k'_{2d}C_S^2 + \cdots + k'_{md}C_S^m \qquad (4.17)$$

In Equation (4.17) the various modes of decay due to site poisoning, diffusion, and so on, are approximated by the various power-law terms in the series, and it is assumed that the several decay reactions proceed in parallel.

The major simplifying assumption in Equation (4.17) is the implicit assumption that decay is not a function of reactant or product concentration. This is obviously impossible in practice, except in the case where the reactants and products are equally poisonous to the catalyst and both conversion and decay are first order [34,35]. However, even in much more complex cases [34], the time-on-stream function derived from Equation (4.17) describes the observed results very well. The reasons for this are, first, that in practice both reactants and products do contribute significantly to coke formation (see Table 4.1), and second, that Equation (4.17) reduces to a very flexible form on further analysis [33].

We begin by rewriting Equation (4.17) in terms of $\Theta = C_S/C_{SO}$, the fraction of sites still active.

$$- \frac{d\Theta}{dt} = k_{0d} + k_{1d}\Theta + k_{2d}\Theta^2 + \cdots + k_{md}\Theta^m \qquad (4.18)$$

where $k_{md} = k'_{md}C_{SO}^{m-1}$. It has been shown that Equation (4.18) is very well represented by the simpler equation

$$- \frac{d\Theta}{dt} = k_d\Theta^m \qquad (4.19)$$

if the exponent m is allowed to assume nonintegral values [33].
In fact, Equation (4.19) fits data fitted by all the various time-
on-stream equations (4.9) to (4.16) and provides a broadly applicable functional description of catalyst decay.

On integration of Equation (4.19), we obtain

$$\Theta = \left(\frac{1}{1 + Gt}\right)^N \qquad m \neq 1 \qquad (4.20)$$

where $G = (m - 1)k_d t$ and $N = 1/(m - 1)$.

The functional form shown in Equation (4.20) has been successfully applied to a great variety of cracking and dehydrogenation reactions [36—41] and has been the basis of a considerable
improvement in understanding the behavior of cracking catalysts.
In particular, it has allowed an improved understanding of
selectivity behavior on rapidly decaying catalysts.

4.2 SELECTIVITY PHENOMENA ON DECAYING CATALYSTS

The behavior of selectivity in the presence of rapid catalyst decay is far from being intuitively obvious. Because of this, many
misconceptions have arisen with regard to selectivity in catalytic
cracking, and not a few of them persist. To understand the problem more fully, one must appreciate two basic facts: first, the
observed cracking selectivity at a fixed conversion is a function
of reactor type; and second, on confined-bed reactors, the observed selectivity at a fixed conversion is a function of the space
velocity and catalyst/oil ratio used [1,42—44]. These observations are true regardless of whether or not catalyst decay is
selective with respect to the sites responsible for the various reactions that take part in catalytic cracking.

4.2.1 Selectivity Phenomena in Experimental
Confined-Bed Reactors

The phenomena associated with nonselective decay in static-bed
reactors have been described mathematically using TOS models
and applied to gas-oil cracking data [45,46]. A review by Ko
and Wojciechowski [47] describes in detail the morphologies that

appear in selectivity behavior for primary and secondary, stable
and unstable products of catalytic cracking in a fixed-bed reactor. Here we concentrate on the selectivity of a simple series
mechanism

$$A \longrightarrow B \longrightarrow C$$

to illustrate its behavior when this reaction takes place on a
static bed of decaying catalyst.

If catalyst decay is slow in relation to the contact time of feed
in the reactor, each increment of feed traverses the bed seeing
catalyst of essentially the same activity. The conversion of this
increment depends on the activity of the catalyst. Subsequent
increments will exhibit successively lower conversions as they
pass over a more and more decayed catalyst. If we concentrate
for the moment on the still simpler mechanism

$$A \longrightarrow B$$

we can write for any given increment of feed

$$\frac{dC_A}{d\tau} = k'C_S C_A \tag{4.21}$$

and hence

$$\frac{C_A}{C_{A0}} = 1 - X_A = \exp(-k'C_S \tau) \tag{4.22}$$

where τ is the space time and C_S is the concentration of sites
active during the passage of the increment in question. Equation
(4.22) gives the instantaneous fraction of feed unconverted. To
obtain the average fraction of feed converted, we average over
all increments that pass over the catalyst in the time-on-stream
period from 0 to t_f.

$$\overline{X}_A = \frac{1}{t_f} \int_0^{t_f} X_A \, dt \qquad (4.23)$$

On substitution of Equation (4.20) for C_S / C_{S0} in (4.22) we arrive at

$$\overline{X}_A = 1 - \frac{1}{t_f} \int_0^{t_f} \exp[-k_0(1 + Gt)^{-N} \tau] \, dt \qquad (4.24)$$

where $k_0 = k' C_{S0}$ and all else as before. In Equation (4.24) two "times" appear; t, the time on stream, and τ, the contact time. These have been shown to be related in the following way [48]:

$$\tau = bPt_f \qquad (4.25)$$

where b is a constant, t_f the duration of the run, and P the catalyst-to-reagent ratio in the run. Thus we write

$$\overline{X}_A = 1 - \frac{1}{t_f} \int_0^{t_f} \exp[-k_0(1 + Gt)^{-N}(bPt_f)] \, dt$$

$$= f(k_0, G, N, b, P, t_f) \qquad (4.26)$$

In Equation (4.26) k_0, G, N, and b are fixed parameters associated with the system under study, while P and t_f are experimental variables.

To study the reaction we must vary both P and t_f or, using Equation (4.25), any two of the variables τ, P, and t_f. The preferred experimental procedure is to fix P and vary τ (and consequently t_f) over some range. Then a new P is chosen and the variation of τ repeated.

In such a series of experiments at a constant catalyst-to-oil ratio, regardless of the decay behavior of the system, we will find that $\overline{X}_A = \overline{X}_B$ for the simple mechanism being considered. A plot of $\overline{X}_B$ versus $1 - \overline{X}_A$ will be a straight line, a fact that seems intuitively obvious. Things become much less obvious when we return to our original system of consecutive reactions:

$$A \xrightarrow{\ k_1\ } B \xrightarrow{\ k_2\ } C$$

The instantaneous yields and conversion for such a system are well known and widely understood in their behavior.

$$\frac{C_A}{C_{A0}} = 1 - X_A = \exp(-k_{10}C_{S1}\tau) \tag{4.27}$$

$$\frac{C_B}{C_{A0}} = X_B = \frac{k_{10}C_{S1}}{k_{20}C_{S2} + k_{10}C_{S1}}$$

$$[\exp(-k_{10}C_{S1}\ \tau) - \exp(-k_{20}C_{S2}\ \tau)] \tag{4.28}$$

$$\frac{C_C}{C_{A0}} = X_C = 1 - X_A - X_B \tag{4.29}$$

When we substitute Equation (4.20) for C_S and come to integrate the instantaneous quantities given in Equations (4.27) to (4.29) with respect to t, we first have to consider the possibility that C_{S1}, the concentration of sites responsible for reaction 1, is not the same as C_{S2}. In addition, there exists the possibility that the two types of sites do not decay with the same kinetics and hence

$$\Theta_1 = (1 + G_1 t)^{-N_1} \neq \Theta_2 = (1 + G_2 t)^{-N_2} \tag{4.30}$$

Such an assumption leads to

$$\overline{X}_B = f_1(k_{10}, k_{20}, G_1, G_2, N_1, N_2, b, P, t_f) \tag{4.31}$$

$$\overline{X}_C = f_2(k_{10}, k_{20}, G_1, G_2, N_1, N_2, b, P, t_f) \qquad (4.32)$$

Simulation by computer has shown that such "selective decay" can not be distinguished visually in static-bed results from selectivity patterns obtained under the assumption of "nonselective decay" (i.e., $G_1 = G_2$, $N_1 = N_2$). For the present we assume nonselective decay.

Plots of $\overline{X}_B$ and $\overline{X}_C$ versus $1 - \overline{X}_A$ are shown in Figures 4.1 and 4.2 for the cases $N > 1$ (class I), $N = 1$ (class II), and $N < 1$ (class III). A number of important observations can be made regarding these two figures. To reduce the discussion to its essentials we will discuss only class III behavior.

First, we note that there is an enveloping curve that overlies all the constant P loops and is denoted as the optimum performance

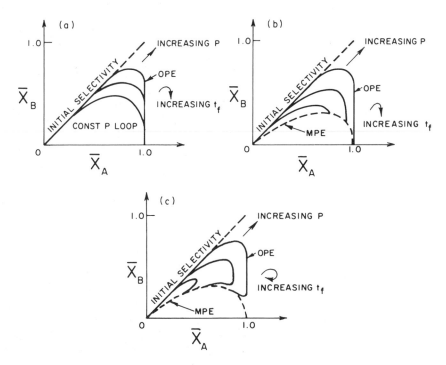

Figure 4.1   Theoretical selectivity plots of unstable primary product B in the reaction A → B → C for a catalyst of (a) class I, (b) class II, and (c) class III. (From Ref. 47.)

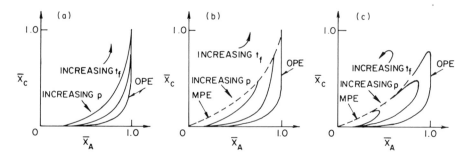

Figure 4.2   Theoretical selectivity plots of stable secondary product C in the reaction A → B → C for a catalyst of (a) class I, (b) class II, and (c) class III.   (From Ref. 47.)

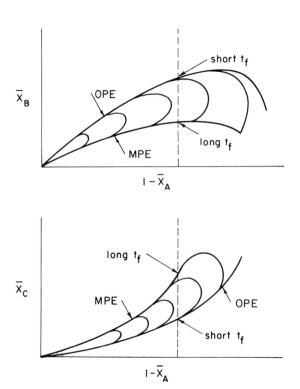

Figure 4.3   Selectivity in a class III consecutive reaction of the type A → B → C.

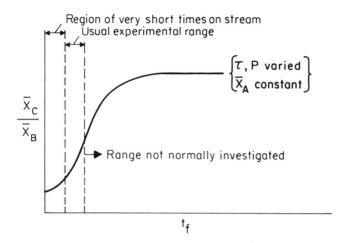

Figure 4.4 Influence of the final TOS on the ratio of product yields at constant conversion in the reaction A → B → C.

envelope (OPE). We will discuss this curve in more detail later; here we merely note that it represents one limit of experimentally accessible behavior on both the $\overline{X}_B$ and $\overline{X}_C$ plot. A second curve, denoted the minimum performance envelope (MPE), defines a second limit of the accessible selectivity behavior. Between these two curves lies an area that is fully accessible to experiment, with the constraint that each point in this area represents the results of a specific and unique set of reaction conditions. No two sets of reaction conditions will give the same point in the area between the OPE and the MPE.

If we concentrate on points lying on a line at constant conversion, we realize that various selectivities can be produced at a given conversion by the appropriate choice of reaction conditions. For example, at very short time on stream the selectivities will lie on the OPE for both $\overline{X}_B$ and $\overline{X}_C$, as shown in Figure 4.3, whereas at long times on stream they will approach the MPE for both. The ratio $\overline{X}_C/\overline{X}_B$ will therefore change with t_f. In fact, if we devise a series of experiments where P and τ are varied so as to keep $\overline{X}_A$ constant and then plot the observed $\overline{X}_C/\overline{X}_B$, we will find the behavior shown in Figure 4.4.

The obvious increase in selectivity for $\overline{X}_C$ in comparison to $\overline{X}_B$ is purely a function of the behavior of Equations (4.31) and (4.32), even though $G_1 = G_2$ and $N_1 = N_2$ and hence no selective loss of sites for one reaction or the other is taking place.

Nonetheless, the observations illustrated in Figure 4.4 can easily
be misinterpreted as proof of the selective poisoning of B-pro-
ducing sites as a function of catalyst age. Other, more complicated types of behavior can be seen by considering other figures in the work of Ko and Wojciechowski [47] dealing with more
complex networks of reactions. All figures in that work were
constructed using the assumption of nonselective poisoning.

The conclusion that emerges from these considerations is that
static-bed and confined fluidized-bed reactors are not capable of
producing results which will unambiguously demonstrate selective
poisoning. What seems at first sight to be clear evidence of
change in catalyst selectivity with time on stream turns out to be
a mirage caused by the unfamiliar behavior of the integral reactor
system. In interpreting real experimental data that show behavior
similar to that shown in Figure 4.1, it has never been necessary
to postulate selective decay in order to obtain a satisfactory con-
gruence between the observed results and equations such as
(4.27) and (4.28) [38,39,45,46,49].

4.2.2   Selectivity in Steady-State and Commercial Reactors

The behavior of experimental selectivity data observed in confined-
bed reactors is not repeated in continuous reactors, be they of
the fluidized- or moving-bed type. This has led to many a dis-
agreement between experimentalists and operating personnel in
refineries.

To make the point clear, let us first consider an ideal fluidized
bed with perfect backmixing of catalyst and plug flow of reactants.
In such a reactor the residence-time distribution for the catalyst
is

$$E(t) = \frac{1}{\bar{t}} \exp\left(-\frac{t}{\bar{t}}\right) \tag{4.33}$$

where t is the time on stream of the individual particle of catalyst,
whereas $\bar{t}$ is the average residence time of all particles and corresponds to t_f in Equation (4.25). At the same time, each particle
that has been on stream for time t has a rate constant given by

$$k_0 \Theta_t = k_0 (1 + Gt)^{-M} \tag{4.34}$$

In this text we call k_0 the "initial rate constant" and $k_0\Theta_t$ the "rate constant" at any other time t. The average rate constant for the whole bed is therefore

$$\bar{k} = \int_0^\infty k_0 (1 + Gt)^{-M} \frac{1}{\bar{t}} \exp\left(-\frac{t}{\bar{t}}\right) dt$$

$$= \frac{k_0}{\bar{t}} \int_0^\infty \frac{\exp(-t/\bar{t})}{(1 + Gt)^M} dt \tag{4.35}$$

If deactivation is nonselective, each rate constant in a reaction network will be modified by exactly the same function as that seen in Equation (4.35). In that case the average yield of $\bar{X}_B$ and $\bar{X}_C$ for the mechanism considered in Section 4.2.1 will be that given by Equations (4.27) to (4.29) with each kC_S replaced by a corresponding $\bar{k}$.

$$\bar{X}_A = \exp(-\bar{k}_1 \tau) \tag{4.36}$$

$$\bar{X}_B = \frac{k_{10}}{k_{20} - k_{10}} [\exp(-\bar{k}_1 \tau) - \exp(-\bar{k}_2 \tau)] \tag{4.37}$$

$$\bar{X}_C = 1 - \bar{X}_A - \bar{X}_B \tag{4.38}$$

It can be shown that points generated by Equations (4.36) to (4.38) fall exactly on the OPE of points generated by Equations (4.27) to (4.29). That is, an ideal fluidized bed always operates at the best selectivity possible for that system [46]. The ideal moving-bed system can also be shown to operate on the OPE as long as decay is nonselective and both catalyst and feed are in perfect plug flow [46].

As soon as imperfections in mixing are introduced into the models described above, they depart in selectivity behavior from the OPE and begin to fill the area available between the OPE and the MPE. The bigger the imperfections, the more of this area is filled.

Since fluidized beds that approach the ideal model assumed here are difficult to operate over an extended range of conditions, let us concentrate on a moving-bed experiment where plug flow in both reactants and products is more readily achievable.

In such a reactor, by changing P and τ while keeping $\bar{t}$ (= t_f) constant, one can generate the selectivity curve at that t over a range of conversions. If the decay of catalyst is nonselective, this selectivity curve will coincide with the OPE of static-bed experiments, regardless of the value of $\bar{t}$. If decay is selective, the moving-bed selectivity curve will be different for each $\bar{t}$. Thus studies of selectivity in moving-bed reactors hold promise of showing whether or not the decay of a given cracking catalyst is selective.

The difference in selectivity behavior between confined-bed and steady-state reactors is so great that it has led to much confusion. Fortunately, most laboratory tests of cracking catalyst performance are now carried out at very short times on stream, thereby approaching the OPE in selectivity. This provides a more realistic evaluation of catalysts for commercial purposes but does not allow a full understanding of catalyst behavior. The result is that catalyst evaluation has been improved, although the intuitive grasp of the effects of decay is not as well developed as it might be.

4.2.3   Influence of Catalyst Properties on Selectivity

Catalyst selectivity, even in the absence of decay, is influenced by a variety of parameters. The simplest is the effect of increased temperature, which increases the rate of reactions with high activation energies faster than that of reactions with low activation energy.

The second well-known effect is that of increased concentration of reactants, which increases the rates of high-order reactions over those of a lower kinetic order. Other concentration effects have been studied and are well known in the literature [34].

If a catalyst is used to carry out a reaction that involves a number of concurrent steps or parallel reactions, selectivity may

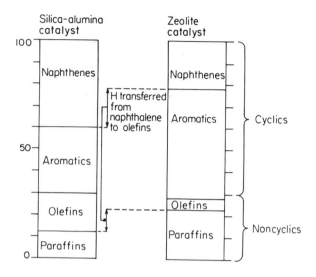

Figure 4.5 Comparison of product distributions in gas-oil cracking catalyzed by silica alumina and Y zeolite. (From Ref. 52.)

depend on the number and nature of the various active sites on the catalyst. This seems to be the case in cracking, where various reactions seem to proceed on different sites [49,50]. As a consequence, various catalyst preparations produce different selectivities from the same feedstock [49—54]. In fact, much of the progress in cracking catalyst development is due to improvements in selectivity behavior. For example, Weisz [51] reports the differences shown in Figure 4.5 between zeolite and silica-alumina catalysts in gas-oil cracking. Since the products shown in Figure 4.5 are produced by primary and secondary reactions from a great variety of feed molecules, it is difficult to tell exactly what is the cause of the observed selectivity differences, but there is no doubt that a significant difference exists between the two catalysts and that it is not caused by temperature or feed concentration effects.

The differences in selectivity displayed by various catalyst preparations can probably be ascribed to effects discussed in Chapter 3. There it was pointed out that various catalyst preparations have different Lewis/Brǿnsted acid site ratios and different acid strength distributions within each type of site. If the various simultaneous reactions proceed on specific acid types and on limited ranges of acidities of these sites, it is not surprising that the selectivities which result are very dependent on the combination of sites present on a catalyst. Very little systematic work has been done in this area [55].

A complicating factor in selectivity studies is introduced by the sieving effect of the pore size in zeolite catalysts. This limits access of reactants to the interior of the crystallite (Table 4.2) and, perhaps more important, "cages" product molecules, forcing them to undergo repeated reactions before exiting the zeolite crystal.

Thus two types of diffusion affect selectivity in different ways: sieving of feed molecules at the entrance to the interior of the crystal results in selection of which feed molecules are cracked and which are not [57]; and caging of product molecules results in the enhancement of secondary reactions that follow the initial cracking event. These secondary reactions can influence the octane of the gasoline produced by inducing branching and aromatization and can control coke make, the olefin-to-paraffin ratio, and the dry gas make. Figure 4.6 shows the magnitude of diffusivity changes induced by relatively small changes in molecular diameter.

The Lewis/Brønsted site ratio is also though to contribute to the control of the olefin-to-paraffin ratio, coke formation, and hydrogen transfer [58,59]. What proportion of a given effect is due to what cause is far from being understood.

Spatial distribution of sites on the internal surface of the catalyst crystal may also have a significant influence on the reaction. For example, the catalytic coke formed in various zeolites may require the presence of at least two sites at a suitable distance from each other [60]. Furthermore, complicated reactions such as cyclization or dehydrogenation may be "structure dependent" [61, 62] and require very specific pore and site configurations.

Over and above the preceding selectivity-controlling factors, we have the complication of decay, selective or not. If catalyst decay is nonselective, the only consequence of the change of activity is that observed only in confined-bed reactors, and hence only in the laboratory. If, however, decay is selective or induces progressive constriction of pores and hence diffusion limitations, things get very much more complicated. Deactivation complicated by diffusion has been studied by several authors [62–66]. Wheeler's work stands out as the seminal approach. Subsequent work in this area has largely concentrated on the question of activity loss, and much needs to be done to understand selectivity changes induced by increasing diffusional resistance in the presence of a complicated feed. Experimental work has shown that preexisting diffusional control in a catalyst that is cracking gas oil affects the

Table 4.2 Effect of Molecular Diameter on Reaction Rate in the Presence of Large Pores (Si-Al) and Micropores (ReHX Zeolite)

| Reactant Hydrocarbon | SiO_2-Al_2O_3 | | REHX | | Ratio |
	Rate constant	C on catalyst[a] (%)	Rate constant	C on catalyst[a] (%)	$k_{REHX}/k_{SiO_2-Al_2O_3}$
n-Hexadecane	60	0.1	1,000	1.4	17
1,3,5 Triethyl benzene	140	0.4	2,370	2.0	17
2,3,6 Trimethyl-naphthalene	190	0.2	2,420	0.7	13
Phenanthrene	205	0.2	953	1.0	4.7
Analene	210	0.4	513	1.6	2.4

[a]Coke on catalyst after the 2-min test.
Source: Ref. 56.

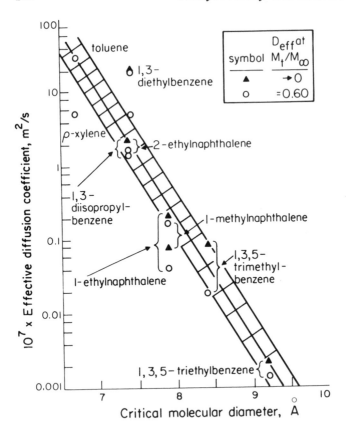

Figure 4.6 Dependence of effective diffusion coefficient on critical molecular diameter of the diffusing species; adsorptive counterdiffusion into cyclohexane-saturated NaY at 25°C, free pore aperture = 7.4 Å. M_t/M_∞ denoted the fractional approach to equilibrium. (From Ref. 57.)

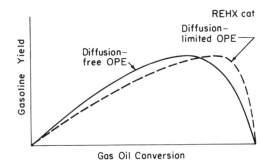

Figure 4.7   Effect of diffusion on gasoline selectivity. (From Ref. 67.)

apparent order of the main cracking reaction and reduces its rate much more than it affects gasoline recracking [67]. This shifts maximum gasoline yield to higher conversions, as shown in Figure 4.7.

In principle, therefore, if catalyst aging simply induces diffusion limitations in gas-oil cracking in steady-state reactors, one might expect a drift from the diffusion-free OPE to the diffusion-limited OPE with increasing decay. This phenomenon is very similar to that which would be observed in a moving bed if selective poisoning of gas-oil cracking sites were taking place, adding to the difficulties of interpretation.

In conclusion, it is safe to say that neither the exact mechanism nor the chemistry of catalyst decay is known. Fortunately, the TOS theory, although known to be a simulation of more complex realities, allows an excellent fit to experimental data from cracking reactions over broad ranges of conditions, using both pure and mixed feeds. At this time there is no evidence from such studies that catalyst decay is due to anything but a fairly simple chemical poisoning of individual active sites. There is also no evidence of selective poisoning of sites of the various strengths that take part in the cracking reaction. Some evidence exists that Brønsted and Lewis sites decay by different mechanisms.

In view of the hazards of misinterpretation inherent in this complex system, it will require much careful experimentation and a thorough understanding of the significance of the results to make any more definite statements than the above. This is clearly not a problem that will be resolved by a "quick and dirty" experiment.

REFERENCES

1. J. B. Butt, *Adv. Chem. Ser.* (Chem. React. Eng., Int. Symp., 1st), *109*: 259 (1972); H. H. Lee and J. B. Butt, *AIChEJ.*, *28*: 405 (1982); *28*: 410 (1982); F. S. Kovarik and J. B. Butt, *Catal. Rev. Sci. Eng.*, *24*: 411 (1982); C. C. Lin, S. W. Park, and W. J. Hatcher, Jr., *Ind. Eng. Chem.*, *Process Des. Rev.*, *22*: 609 (1983); M. A. Pacheco and E. E. Petersen, *J. Catal.*, *86*: 75 (1984); *88*: 400 (1984); E. K. Reiff, Jr. and J. R. Kittrell, *Ind. Eng. Chem.*, *Fundam.*, *19*: 126 (1980)

2. E. E. Wolf and F. Alfani, *Cat. Rev. Sci. Eng.*, *24*: 329 (1982)

3. K. J. Laidler, *The Chemical Kinetics of Enzyme Action*, New York, Oxford, 1958

4. R. N. Cimbalo, R. L. Foster, and S. J. Wachtel, *Oil Gas J.*, *70*(20): 112 (1972)

5. E. T. Habib, H. Owen, P. W. Snyder, C. W. Streed, and P. B. Venuto, *Ind. Eng. Chem.*, *Prod. Res. Dev.*, *16*: 291 (1977)

6. R. D. Haldeman and M. C. Botty, *J. Phys. Chem.*, *63*: 489 (1959)

7. H. R. Grane, J. E. Connor, and G. P. Masologites, *Pet. Refiner*, *40*(5): 168 (1961)

8. C. H. Tan and O. M. Fuller, *Can. J. Chem. Eng.*, *48*: 174 (1970)

9. J. B. Butt, S. Delgado-Diaz, and W. E. Muno, *J. Catal.*, *37*: 158 (1975)

10. J. W. Beeckman and G. F. Froment, *Ind. Eng. Chem.*, *Fundam.*, *18*: 245 (1979)

11. K. S. Tsakalis, T. T. Tsotsis, and G. J. Stiegel, *J. Catal.*, *88*: 188 (1984)

12. A. Wheeler, *Catalysis*, Vol. II (P. H. Emmett, Ed.), p. 105, New York, Reinhold, 1955

13. D. D. Do and P. F. Greenfield, *Chem. Eng. J.*, *27*: 99 (1983)

14. G. F. Froment and K. B. Bischoff, *Chem. Eng. Sci.*, *16*: 189 (1961)

15. I. Nam and J. R. Kittrell, *Ind. Eng. Chem. PDD*, *23*: 237 (1984)

16. W. J. Hatcher, Jr., *Ind. Eng. Chem. PRD*, *24*: 10 (1985)

17. B. W. Wojciechowski, *Catal. Rev. Sci. Eng.*, *9*: 79 (1974)

18. V. W. Weekman, Jr., *Ind. Eng. Chem. PDD*, *7*: 90 (1968)

19. G. F. Froment, *Stud. Surf. Sci. Catal.* (Catal. Deact.), *6*: 1 (1980)

20. R. P. DePauw and G. F. Froment, *Chem. Eng. Sci.*, *30*: 789 (1985)

21. J. W. Beeckman and G. F. Froment, *Ind. Eng. Chem. Fundam.*, *18*: 245 (1979)

22. C. J. Plank and D. M. Nace, *Ind. Eng. Chem.*, *47*: 2474 (1955)

23. H. Heinemann, *Ind. Eng. Chem.*, *43*: 2098 (1951)

24. T. M. John, R. A. Pachovsky, and B. W. Wojciechowski, *Adv. Chem. Ser.*, *133*: 422 (1974)

25. E. B. Maxted, *Adv. Catal.*, *3*: 129 (1951)

26. E. F. G. Herington and E. K. Rideal, *Proc. R. Soc. (London)*, *A184*: 434 (1945)

27. V. W. Weekman, Jr., *Ind. Eng. Chem.*, *Process Des. Dev.*, *7*: 90 (1968)

28. A. Corma, A. Lopez Agudo, I. Nebot, and F. Tomas, *J. Catal.*, *77*: 159 (1982)

29. H. V. Maat and L. Moscou, *Proc. 3rd Int. Congr. Catal.*, *Amsterdam*, *2*: 1277 (1964)

30. A. L. Pozzi and H. F. Rase, *Ind. Eng. Chem.*, *50*: 1075 (1958)

31. F. H. Blanding, *Ind. Eng. Chem.*, *45*: 1186 (1953)

32. A. Voorhies, Jr., *Ind. Eng. Chem.*, *37*: 318 (1945)

33. B. W. Wojciechowski, *Can. J. Chem. Eng.*, *46*: 48 (1968)

34. O. Levenspiel, *J. Catal.*, *25*: 265 (1972)

35. M. R. Viner and B. W. Wojciechowski, *Can. J. Chem. Eng.*, *60*: 127 (1982); ibid. (in press)

36. D. R. Campbell and B. W. Wojciechowski, *J. Catal.*, *23*: 307 (1971)

37. R. A. Pachovsky and B. W. Wojciechowski, *Can. J. Chem. Eng.*, *49*: 365 (1971)

38. A. Borodzinski, A. Corma, and B. W. Wojciechowski, *Can. J. Chem. Eng.*, *58*: 219 (1980)

39. T. M. John and B. W. Wojciechowski, *J. Catal.*, *37*: 240 (1975)

40. A. Corma, R. Cid, and A. Lopez Agudo, *Can. J. Chem. Eng.*, *57*: 638 (1979)

41. A. Corma, J. B. Monton, and V. Orchilles, *Ind. Eng. Chem. PRD*, *23*: 404 (1984)

42. A. Romero, J. Bilbao, and J. R. Gonzalez, *Chem. Eng. Sci.*, *36*: 797 (1981)

43. J. R. Gonzalez, M. A. Gutierrez, J. I. Gutierrez, and A. Romero, *Chem. Eng. Sci.*, *39*: 615 (1984)

44. G. J. Frycek and J. B. Butt, *Adv. Chem. Ser.* (Chem. Catal. Reactor Modelling), *237*: 375 (1984)

45. V. W. Weekman, Jr. and D. M. Nace, *AIChEJ*, *16*: 397 (1970)

46. R. A. Pachovsky and B. W. Wojciechowski, *Can. J. Chem. Eng.*, *53*: 308 (1975)

47. A.-N. Ko and B. W. Wojciechowski, *Prog. React. Kinet.* (1984) in press

48. D. R. Campbell and B. W. Wojciechowski, *J. Catal.*, *20*: 217 (1971)

49. A. Corma and B. W. Wojciechowski, *Catal. Rev. Sci. Eng.*, *27*: 29 (1985)

50. A. Corma and B. W. Wojciechowski, *Catal. Rev. Sci. Eng.*, *24*: 1 (1982)

51. P. B. Weisz, *Chem. Technol.*, *3*: 498 (1973)

52. S. C. Eastwood, C. J. Plank, and P. B. Weisz, *Proc. 8th World Pet. Congr., Moscow*, *4*: 245, 1971

53. R. Beaumont, P. Pichat, D. Barthomeuf, and Y. Trambouze, *Proc. 5th Int. Congr. Catal.*, *1*: 343 (1972)

54. D. M. Nace, *Ind. Eng. Chem., Prod. Res. Dev.*, *8*: 31 (1969)

55. A. Corma and B. W. Wojciechowski, *Can. J. Chem. Eng.*, *60*: 11 (1982)

56. D. M. Nace, *Ind. Eng. Chem.*, *Prod. Res. Dev.*, *9*: 203 (1970)

57. R. M. Moore and J. R. Katzer, *AIChEJ*, *18*: 816 (1972)

58. W. K. Hall, F. E. Lutinski, and H. R. Gerberich, *J. Catal.*, *3*: 512 (1964)

59. P. E. Eberly, Jr., C. N. Kimberlin, W. H. Miller, and H. V. Drushel, *Ind. Eng. Chem.*, *Process Des. Dev.*, *5*: 193 (1966)

60. L. D. Rollman and D. E. Walsh, *J. Catal.*, *56*: 139 (1979)

61. C. D. Chang and A. J. Silvestri, *J. Catal.*, *47*: 249 (1977)

62. S. Masamune and J. M. Smith, *AIChEJ*, *12*: 384 (1966)

63. J. J. Carberry and R. L. Gorring, *J. Catal.*, *5*: 529 (1966)

64. S. J. Khang and O. Levenspiel, *Ind. Eng. Chem.*, *Fundam.*, *12*: 185 (1973)

65. L. L. Hegedus and E. E. Petersen, *Catal. Rev. Sci. Eng.*, *9*: 245 (1974)

66. H. W. Haynes, Jr. and K. Leung, *Chem. Eng. Commun.*, *23*: 161 (1983)

67. R. A. Pachovsky and B. W. Wojciechowski, *AIChEJ*, *19*: 1121 (1973)

5

Catalytic Cracking of Pure Compounds

5.1   INTRODUCTION

The early workers in catalytic cracking took up the study of pure
component cracking in the hope of establishing the processes and
mechanisms involved. The initial rush of experimentation dealt
mainly with the cracking of alkanes and concentrated on determining the cracking patterns and relative rates of cracking for a
variety of branched and unbranched alkanes. No account was
taken of catalyst decay, and conclusions were drawn on the basis
of scattered runs at relatively high conversions. Even with these
handicaps, compounded as they were by analytical difficulties,
general ideas were developed that hold to this day. The β-crack-
ing rule was proven, and the relative ease of carbenium ion formation on primary, secondary, and tertiary carbons was established.
    Unfortunately, this type of work was of little practical use and
was soon dropped in industrial laboratories. University laboratories avoided this type of work because the system was "dirty";
the catalyst decayed rapidly and no systematic methods or quantitative treatments were available to deal with the problem. What
work continued was done mainly in industrial laboratories in
search of a standard test for catalyst evaluation. From these
efforts emerged a number of studies on cumene cracking and the
still-used α test. Progress was slow, however, and the treatment of the results was superficial, owing to the continuing lack
of a satisfactory methodology.
    In the 1970s a number of workers began to apply sophisticated
mathematical theories backed by digital computer power for the
necessary quantification of experimental results. At the same
time, theories attempting to deal with catalyst decay began to
emerge in two forms: coke-on-catalyst and time-on-stream
theories. Some of these evolved into forms that can be conveniently applied to industrial and laboratory cracking data.
    This development, together with greatly improved analytical
methods, computational facilities, and well-defined experimental

methodologies, has led to a revival in studies of pure component cracking. The modern studies tend to be done at institutes and universities, where the fundamental nature of the information being obtained is at least "academically acceptable." As time goes on there emerges a real possibility that this type of work will allow a kinetic mapping of active centers on cracking catalysts and permit rational catalyst design.

5.2   CATALYTIC CRACKING OF PARAFFINS

Paraffins are an important component of both the gas-oil feeds used in many refineries and the primary products of cracking produced in all refineries. The many homologous series of paraffinic compounds are admirably suited for systematic studies of the effects of molecular structure on the cracking behavior of catalysts. It comes as no surprise, therefore, that a substantial amount of work was done on paraffin cracking in the early years of cracking research. What is surprising is that the work was not continued or developed in any systematic way and that as a result we are in a thoroughly unsatisfactory state of knowledge regarding these reactions. Much of the early work in this field reports results at high conversions which are contaminated to such an extent by thermal reactions that only the most general conclusions can be drawn.

The field has recently been reactivated by a number of authors [1—5] who have approached the topic armed with modern analytical, computational, and theoretical developments and are duly aware of the importance of accounting for thermal reactions and for catalyst decay.

Much more remains to be done. Here we sketch the early results and introduce the new methodology which is used to obtain fundamentally sound, quantitative information regarding the action of cracking catalysts on paraffinic molecules.

5.2.1   Rate of Reaction and Product Distribution

One of the early favorites in the study of pure component cracking was the cracking of n-hexadecane. Much of the early work was done on molecules of this size because they are representative of the size of molecules found in refinery feedstocks, but the more easily understood cracking of short-chain molecules was also studied in some detail. By studying a series of homologous

Table 5.1   Influence of Chain Length
on the Activation Energy for Cracking

Hydrocarbon	Activation energy (kcal/g·mol)
$n\text{-}C_6H_{14}$	36.6
$n\text{-}C_7H_{16}$	29.4
$n\text{-}C_8H_{18}$	24.9

Source: Ref. 6.

short-chain paraffins it was observed [6] that the length of the chain has an influence on the activation energy of the cracking reaction, as shown in Table 5.1.

In keeping with points raised previously in connection with carbocation formation, it is clear that this observation testifies to the increasing ease of formation of carbocations on the interior carbons of linear molecules.   Earlier, Voge reached a similar conclusion by comparing the conversion of various paraffins over silica-alumina-zirconia at standard conditions and 500°C [7], as shown in Table 5.2.

Good et al. [8] studied the influence of branching on the rate of reaction. They found that the various isomers of hexane cracked at 550°C, as shown in Table 5.3. From such work it became clear that not all carbon-carbon bonds are equally easy to

Table 5.2   Influence of Chain Length
on the Degree of Conversion at
Standard Conditions

Paraffin	Percent conversion
$n\text{-}C_5H_{12}$	1
$n\text{-}C_7H_{16}$	3
$n\text{-}C_{12}H_{26}$	18
$n\text{-}C_{16}H_{34}$	42

Source: Ref. 7.

Table 5.3   Influence of Chain Branching on
Conversion at Standard Conditions

Isomer	Percent conversion
C—C—C—C—C—C	13.8

C—C—C—C—C
  |
  C
	24.9

C—C—C—C—C
    |
    C
	25.4

C—C—C—C
  | |
  C C
	31.7

C
  |
C—C—C—C
  |
  C
	9.9

Source: Ref. 8.

crack. This is of importance in shaping the selectivity of catalysts and agrees with the findings of more recent molecular orbital
calculations [9,10] and with the behavior of carbocations in
solution.

In studying the products of cracking, a question of major importance arises: Which of the many products observed are the
primary products of the reaction? Only recently, as the experimental and theoretical techniques necessary for this type of work
were developed, have systematic efforts been made at identifying
the primary products.

It is most unfortunate that most of the early catalytic cracking
investigations followed the rate of cracking by following the percentage yield of various products. This method of activity interpretation has led to a great deal of confusion when attempting
to correlate the works of various authors. Furthermore, as we
will see later, the question of catalyst decay was largely ignored,
and results were often reported on catalysts at significantly

Table 5.4   Product Distribution in Cracking Hexadecane on
Si-Al-Zr

C_1	C_2	C_3	C_4	C_5	C_6	C_7	C_8	C_9	C_{10}	C_{11}	C_{12}	C_{13}	C_{14}
5	12	97	102	64	50	8	8	3	3	2	2	2	1

Source: Ref. 11.

different times on stream. Catalyst age is now known to lead to
a great deal of variation in the distribution of products (see Section 4.2), making it even more difficult to compare the various
works reported in the literature.

For instance, Greensfelder and Voge [11] report the yields obtained in the cracking of hexadecane at 500°C at 24% conversion
over a catalyst of silica-alumina-zirconia to be those listed in
Table 5.4. We see that for 100 mol of hexadecane at a conversion
of 24%, one obtains 359 mol of hydrocarbons (plus 12 mol of
hydrogen). This distribution is reported to vary only slightly in
the range of conversions between 11 and 68%. However, some of
these products are almost certainly not the primary products of
cracking; indeed, significant proportions may be due to thermal
cracking, which occurs concurrently with the catalytic cracking
process. If one were observing only the primary products of a
catalytic cracking reaction, one would expect to see the following
product mole ratios:

$$\frac{C_1}{C_{15}} = \frac{C_2}{C_{14}} = \frac{C_3}{C_{13}} = \frac{C_4}{C_{12}} = \frac{C_5}{C_{11}} = \cdots = 1$$

Consecutive cracking of the carbocation left after the initial
cracking event would complicate this picture, but only runs at
much lower conversions would tell how much the picture presented
by Table 5.4 is distorted by secondary reactions. Nonetheless,
from such results, Greensfelder et al. [12] proposed that the
formation of carbocations should proceed at approximately the
same rate for all the CH_2 groups, and at a much lower rate for
the CH_3 groups on the ends of the molecules. On the basis of
his observations, Voge proposed that there is a rapid equilibra-
tion of all secondary carbenium ions, with subsequent cracking of
these various ions according to the β rule until the chain lengths

Table 5.5   Predicted Product Distribution for Hexadecane
Cracking

C_1	C_2	C_3	C_4	C_5	C_6	C_7	C_8	C_9	C_{10}	C_{11}	C_{12}	C_{13}	C_{14}
0	0	95	97	72	41	7	6	5	4	4	4	0	0

Source: Ref. 12.

are reduced to species of C_6 or less. Furthermore, the olefins
produced in the reaction should follow similar cracking rules. On
the basis of such suppositions, the products calculated for 100 mol
of hexadecane feed correspond to 359 mol of hydrocarbon with the
distribution shown in Table 5.5. These values were judged to be
in good agreement with those found experimentally, leading to the
conclusion that the proposed mechanism is in fact operating.

The use of mass spectroscopy, together with studies of molecules containing radiocarbon tracers, have greatly improved our
understanding of such reactions. For instance, Emmett and others
[13–15] found that the olefins formed in the primary cracking
reaction undergo a much greater number of parallel and consecutive
reactions than do the product paraffins, which in their short-chain
form are almost completely inert. The olefins whose chain length
is six or more are quickly cracked, whereas C_4 and C_5 olefins
react to form higher-molecular-weight materials, aromatics and
coke. Ethylene and benzene were found to be almost completely
inert.

In another experiment Hightower and Emmett [16] used a mixture
of radioactive propylene and hexadecane as feed on silica alumina at
370°C. It was found that the major part, some 90% of the propylene,
was converted to propane and products in the range C_6 to C_{12}.
Furthermore, at least one-third of the benzene formed came from
propylene. They report the distribution of products to be that
shown in Table 5.6.

A detailed chromatographic investigation of the products allowed
the identification of 60 hydrocarbons in the range C_1 to C_{10}, of
which some 2% were C_6 to C_{10} aromatics. The products were
found to contain 158 mol of olefins and 102 mol of paraffins, while
coke represented 79 mol [14].

Good et al. [8], who studied the cracking of five isomers of
hexane, report that on silica alumina the predominant reaction is
cracking, with minor contributions from reactions such as isomerization, dehydrocyclization, and dehydrogenation. Their distribution

Table 5.6 Product Distribution for Cracking a Mixture of Propylene and Hexadecane

C_1	C_2	C_3	C_4	C_5	C_6	C_7	C_8	C_9	C_{10}	C_{11}	C_{12}	C_{13}	C_{14}
0.7	5.1	96.6	163.1	68.0	18.5	14.1	6.1	1.9	0.5	1.3	2.5	—	—

Source: Ref. 16.

Table 5.7   Product Distribution for Cracking of *n*-Hexane on
Silica Alumina

H_2	CH_4	Total C_2	Total C_3	Total C_4	Total C_5
9.4	12.5	19.5	49.1	8.5	1.0

Source: Ref. 8.

of products in the cracking of *n*-hexane at 550°C and 60 min on
stream is shown in Table 5.7.

Several observations can be made on the basis of this product
distribution. First, there is an unexpectedly large amount of
hydrogen. Second, the high ratio of C_2 to C_4 indicates that
simple cracking is not the only reaction here. Similarly, the
high ratio of C_1 to C_5 confirms that reactions other than simple
cracking are occurring on this catalyst. Indeed, the high levels
of hydrogen, methane, and C_2 products suggest that a significant
amount of thermal cracking took place in this experiment. Furthermore, we are now well aware that it is difficult to say anything about results obtained at a high level of conversion on a
catalyst that has been on stream for 60 min prior to taking the
sample. In fact, after 60 min on stream the catalyst is likely to
be thoroughly deactivated; under such conditions it is difficult to
correlate any instantaneous result with the inherent properties of
the catalyst.

More recently, Pickert et al. [17] have studied the cracking of
n-hexane on zeolites X and Y exchanged with several divalent ions.
They found that MgY, which is the most active of their reported
catalysts, gives some 5% cracking and 16% isomerization at 363°C
at their conditions. With silica alumina they report that a temperature of 475°C was required to achieve the same level of conversion and then only 1.5% of isomerization was obtained. Com-
parisons of these two results can be highly misleading because
they introduce temperature as a variable, making it difficult to
distinguish catalytic effects from temperature effects. The various
products are almost certainly formed by processes requiring different activation energies. Indeed, it has been shown [18] that
on a calcium Y zeolite the cracking of *n*-hexane increases with
temperature from 300°C to 550°C, while isomerization goes through
a maximum at 350°C.

Miale et al. [19] report that HY, REY, and H mordenite are
about 10,000 times more active than amorphous silica alumina for

Table 5.8   Product Distribution for Cracking n-Hexane on HY (mol %)

CH_4	C_2H_6	C_2H_4	C_3H_8	C_3H_6	i-C_4H_{10}	n-C_4H_{10}	C_4H_8	C_5's
0.6	2.2	0.2	47.1	6.8	17.0	7.4	1.1	17.6

Source: Ref. 19.

cracking of n-hexane. Table 5.8 shows the products they obtained with HY zeolite.

Once again such results are difficult to interpret because of the experimental techniques used. However, several facts are worth pointing out. First, there is a high ratio of paraffin to olefin in the product. Second, there is a high ratio of C_4 to C_2 and of C_5 to C_1. One therefore has to suppose that reactions other than cracking are occurring in this system. The high ratio of paraffin to olefin is ascribed to the saturation of olefins via hydrogen-transfer reactions. High ratios of C_4 to C_2 and C_5 to C_1 cannot be explained by simple cracking, and since initial selectivity values are not available, it is difficult to discuss the matter further. One explanation which comes to mind is that perhaps disproportionation or polymerization reactions followed by cracking are taking place.

Schulz and Geertsema [20] have compared a 25% silica alumina to a commercial zeolite catalyst (CBZ-1) in the cracking of n-dodecane. Their results show that the product selectivities observed on the two catalysts differ because of differences in activity as well as because of differences in the rates of decay.

On the basis of the evidence presented thus far, one can conclude the following. The reported selectivity of various catalysts is significantly dependent on the conversion and temperature at which the observations are made and on the catalyst activity. Furthermore, it seems obvious that the product distribution must be a function of the nature of the acid sites, their strength distribution, and the crystalline structure of the material used as catalyst. All these effects are important because they contribute to the variety of reactions that can occur on a given catalyst. For example, various parallel reactions may occur on sites with different acid strength. Indeed, it has been found that mordenites are better catalysts for hydrogen-transfer reactions than either Y zeolites or amorphous silica alumina, and consequently a higher ratio of paraffin to olefin products is formed by cracking on mordenites [21,22].

Using the same catalyst and consequently the same crystalline structure, various conversions and product distributions can be obtained by changing the total number and the strength distribution of the acid sites available. By changing pretreatments, for example, both the number and strength distribution of the acid sites can be changed. The activity for cracking *n*-hexane or *n*-heptane on zeolites is changed by such pretreatments [23−25]. The level of exchange as well as the type of cation introduced into the zeolite also has an important influence on the number and strength of acid sites and consequently on the final activity and selectivity of the catalyst. Since there is a possibility that not all acid sites decay at the same rate, catalyst decay may also influence the selectivity of a given catalyst. The obvious conclusion is that unless a very well defined and systematic experimental program is undertaken, very little reliable information can be extracted from a given set of experiments.

5.2.2   Recent Studies of Rate of Reaction and Product Distribution

Recently, an extensive study of the cracking of *n*-heptane on chromium-hydrogen-sodium Y zeolites was carried out in the range 400 to 470°C [2−4]. In this work the selectivity curves for the various products observed were obtained in the absence of deactivation by using optimum performance envelope curves [26] as described in Section 4.2, and shown in Figure 5.1. In this way all the observed products were classified as primary, secondary, stable, or unstable and their initial selectivities were measured from the plots.

By observing the values of the uncorrected initial selectivities reported in Table 5.9, one can see that the ratios of C_5 to C_2 and C_4 to C_3 and of the paraffins to olefins are higher than 1, especially at the lower temperatures. Furthermore, the selectivities do not add up to 1, as they should if all primary processes in the mechanism are properly accounted for.

It was found that the C_6 fraction is a primary product in this reaction, whereas C_1 is a secondary product and hence C_6 is not produced by the direct cracking of the C_7 feed molecule. Since the minor aromatic products that were found are secondary, no direct dehydrocyclization of *n*-heptane takes place.

The appearance of C_6 as a primary product, as well as the anomalous initial selectivity ratios of C_5 to C_2 and C_4 to C_3, can be explained by assuming that C_6, C_5, and C_4 are in part formed by the disproportionation of two feed molecules.

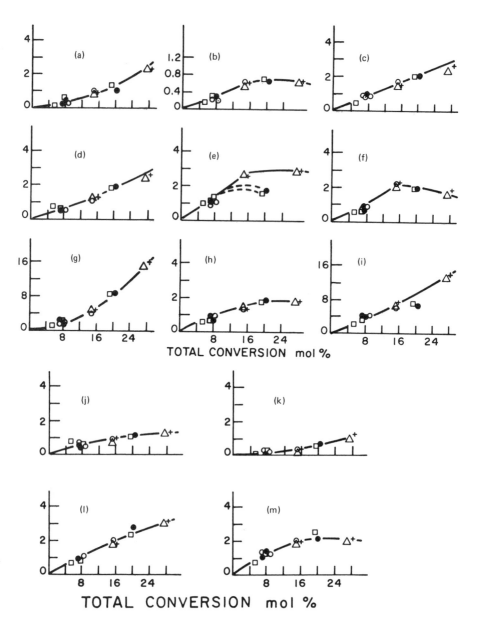

Figure 5.1　Selectivity plots for the products formed during the cracking of n-heptane on a CrHNaY zeolite catalyst:　(a) methane; (b) ethylene; (c) n-butane; (d) ethane; (e) propylene; (f) butenes; (g) propane; (h) C_5 fraction; (i) i-butane; (j) C_5 monocyclic; (k) aromatic; (l) C_6 cyclic; (m) (2 + 3)-methylcyclohexane.　(From Ref. 2.)

138

Table 5.9 Uncorrected Initial Selectivities of the Primary Reaction Products of *n*-Heptane Cracking on CrHNaY

Product	Type	Initial selectivity		
		400°C	450°C	470°C
Ethane	Primary + secondary stable	0.06	0.09	0.12
Ethylene	Primary + secondary unstable	0.03	0.05	0.11
Propane	Primary + secondary stable	0.27	0.27	0.19
Propylene	Primary + secondary stable	0.14	0.21	0.21
n-Butane	Primary + secondary unstable	0.13	0.15	0.12
i-Butane	Primary + secondary stable	0.40	0.35	0.27
Butenes	Primary + secondary unstable	0.09	0.18	0.24
C_5 fraction	Primary unstable	0.13	0.17	0.24
C_6 fraction	Primary unstable	0.16	0.15	0.13
(2 + 3) methylhexane	Primary unstable	0.17	0.11	0.01
	Total	1.58	1.73	1.64

Source: Ref. 2.

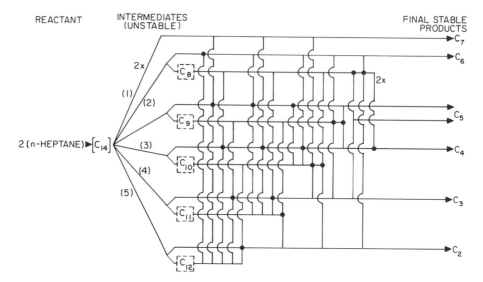

Figure 5.2   Network of potential indirect cracking reactions in
n-heptane cracking.   (From Ref. 2.)

Such disproportionation of paraffins has been observed over
other acid catalysts [27].   When such reactions are taken into
account in the cracking scheme and the initial selectivities are
corrected for their effects, the ratios of C_5 to C_2 and C_4 to C_3
approach 1 within the limits of experimental accuracy, as shown
in Table 5.10.   The complex reaction network shown in Figure
5.2 makes direct experimental measurement of real initial selectivities impossible.

To correct the observed selectivities, certain reasonable assumptions have to be made.   First, it seems clear that some of
the cracking takes place via the dimerization-cracking route, as
no C_1 is produced initially whereas C_6 is a major initial product
in the reaction.   The authors [2] have assumed that both direct
and indirect cracking occurs and that reactions 1 and 2 in Figure 5.2 are much faster than 3, 4, or 5.   This leads to the following scheme of important indirect processes:

Table 5.10 Corrected Initial Selectivities of the Primary Reaction Products of *n*-Heptane Cracking on CrHNaY

Product		Initial selectivity		
		400°C	450°C	470°C
C_2	Cracking	0.09	0.14	0.23
C_3	Cracking	0.37	0.45	0.39
C_4	Cracking	0.38	0.44	0.39
C_5	Cracking	0.09	0.14	0.23
Total for cracking		0.47	0.58	0.62
C_6	Disproportionation	0.16	0.15	0.13
Total for disproportionation		0.32	0.30	0.26
C_7	Isomerization	0.17	0.11	0.09
Total for isomerization		0.17	0.11	0.09
Total corrected selectivities		0.96	0.99	0.97

Source: Ref. 2.

$$2C_7 \longrightarrow [C_{14}] \xrightarrow{\ 6\ } 2C_7$$

$$\xrightarrow{\ 7\ } C_6 + 2C_4$$

$$\xrightarrow{\ 8\ } C_6 + C_5 + C_3$$

$$\xrightarrow{\ 9\ } 2C_6 + C_2$$

In this simplified scheme, only products that are found by experiment to be present initially are considered, and hence no
reactions producing C_8 or higher hydrocarbons are considered.
Next we note that C_2 is produced in the smallest yield in comparison to C_3 and C_4 in Table 5.9. This indicates that reaction
9 is, at best, slow with respect to 6, 7, and 8 and we can take
it to have zero rate. The only remaining disproportionation
reactions are 7 and 8. Reaction 6 simply regenerates the original
reactants.

If one makes the foregoing assumptions, we see that C_2 is a
product of direct cracking. Hence the difference in uncorrected
selectivity between C_5 and C_2 must represent the selectivity for
C_5 via reaction 8. This implies an equal contribution to C_6 and
C_3 selectivity by reaction 8. As no C_1 is produced by initial
processes, by subtracting the selectivity for C_6 by reaction 8
from the total observed selectivity for C_6 we are left with the C_6
selectivity by reaction 7, which is half the C_4 selectivity by
reaction 7. By simple arithmetic one can now calculate the initial
selectivities for C_3 and C_4 by direct cracking of C_7. In this way
the corrected values shown in Table 5.10 were calculated. Figure 5.3 shows the simplified network postulated.

The same calculations allow an estimation of the amount of each
product that arises by mono- and bimolecular processes. From

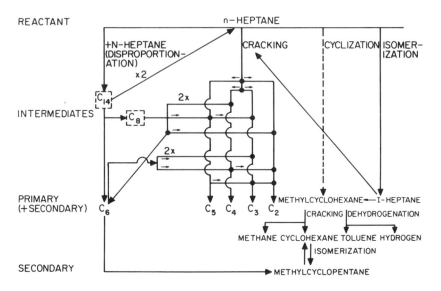

Figure 5.3   Network of the reactions involved in the cracking of
n-heptane. (From Ref. 2.)

Table 5.11   Selectivities for Various Products of n-Heptane
Cracking by the Direct and Indirect Routes

	Corrected initial selectivity					
	400°C		450°C		470°C	
Product	Direct	Indirect	Direct	Indirect	Direct	Indirect
C_2	0.09	0.0	0.14	0.0	0.23	0.0
C_3	0.37	0.04	0.45	0.03	0.39	0.01
C_4	0.38	0.24	0.44	0.24	0.39	0.25
C_5	0.09	0.04	0.14	0.03	0.23	0.01
C_6	0.0	0.16	0.0	0.15	0.0	0.13

Table 5.11 we see that the bimolecular process is dominant for C_6
production and insignificant in C_2 and C_3 production. Further
analysis of the selectivities in Table 5.11 shows that the total
selectivity for monomolecular (direct) cracking of C_7 is 47, 58,
and 62% at 400, 450, and 470°C, respectively. At the same time
the selectivity for conversion of C_7 by bimolecular (indirect)
cracking is 32, 30, and 26%.* The conclusion is therefore that
the higher the reaction temperature, the more the cracking proceeds by the direct monomolecular route. To generalize this concept further, one can say that the shorter the paraffin chain,
the higher the temperature at which it will crack directly by a
monomolecular process. Since the residual selectivity is for
skeletal rearrangement, we see that this is 21, 12, and 11% at the
three temperatures and is clearly becoming less as temperature
increases.

The high ratio of paraffin to olefin in this case is probably
affected by hydrogen-transfer reactions. In Table 5.9 one can
see that this ratio approaches 1 as the temperature of reaction
increases. This is in agreement with the observations of other
authors [28] and suggests that the rate of hydrogen transfer
has a lower activation energy than the rate of desorption.

Clearly, the cracking of paraffins involves various mono- and
bimolecular primary reactions, so that the total disappearance of
feed or of reactant species cannot be used to measure the rate

---

*The total direct cracking selectivity is the sum of $(C_2 + C_3)$ or
$(C_2 + C_4)$ selectivities. The total indirect cracking selectivity is
the sum of $(1/2\ C_4 + C_5 + C_6)$ selectivities.

of cracking, as is commonly done in the literature [29,30]. It is known that the primary reactions taking place in any such system include isomerization and disproportionation as well as cracking itself, while the most important secondary reactions are cracking, hydrogen transfer, and cyclization.

Once this type of information is collected on a given reaction, one can establish the network of reactions that occur in the initial stages of the process. The details of the methodology for this have been explained in reviews [31,32]. Using such methods it is possible to calculate the kinetic rate constants $k_S[S_0]$ for all the primary reactions.

The rate constants calculated in this way include two terms: the specific rate constant k_S, containing a preexponential term and an activation energy term; and a term for the concentration of active sites $[S_0]$. If the activation energies for a given reaction on two catalysts are the same, the corresponding specific rate constants k_S can be taken to be the same, and the ratio of the constants $k_S[S_0]$ can be taken to be the ratio of the active site concentrations. For instance, on CrHNa$_Y$ the rate constant does not increase linearly with the degree of exchange [4]. On the other hand, the nature of the active sites does not change with the level of exchange, as shown by the fact that the activation energy for all reactions remains constant regardless of the level of exchange.

Thus it can be concluded that not all exchanged sites lead to the formation of acid sites that take part in the reactions under study. To find out which fraction of the acid sites is active, the kinetic rate constants for different levels of exchange have been correlated with the number of acid sites of various strengths as calculated by pyridine adsorption-desorption and by titration with butyl amine. From this it was concluded that the cracking of normal heptane occurs only on the stronger acid sites, those with pK values less than +1.5 [4]. A similar conclusion for hexane and isooctane cracking has been reported [33,34].

5.2.3   Effects of Shape Selectivity

The effect of "shape selectivity" on the cracking of straight-chain hydrocarbons is a relatively recently discovered phenomenon. A number of authors [35–40] have reviewed and commented on this phenomenon and on how this factor affects the selectivity of paraffin cracking.

Paraffins can readily enter the pores of zeolites such as mordenite, offretite, and faujasite, and are not significantly constrained by geometric restrictions inside any of them. However, it is well known that in mordenites the main channel system is one-dimensional and that the cations stick out from the channel walls. Partly because of this, as the size of the cation is changed, both the diffusion characteristics and the activity of the mordenite are changed. Wendlant et al. [41,42] have reported results on the cracking of n-octane and methyloctane over H-mordenite catalysts modified by the introduction of alkaline and alkaline earth cations. They discuss the influence of the cation on the basis of its effect on shape selectivity. Yashima et al. [36] have also studied the adsorption and cracking of n-hexane, n-octane, and 3-methylpentane on a series of mordenites whose effective pore dimension was adjusted by the introduction of cations with a suitable ionic radius. These authors found that in the competitive cracking of a mixed feed, BaH mordenite is shape selective against paraffins with quaternary carbon atoms. This effect decreases as the temperature of reaction increases.

In the case of offretite [43], the crystalline structure is formed by channels parallel to the C axis. The pore diameter is 6.3 Å, which allows methylated hydrocarbons such as iso- and neoparaffins to enter [44,45]. When offretite is exchanged by protons, it behaves similarly to faujasite from the point of view of activity and selectivity for paraffin cracking [44,46]. Recently, it has been reported [47] that in offretite the cations located in the gmelinite cage window block the cage entrance, and it is therefore expected that the diffusion of reactant molecules through the gmelinite cavities will depend on the level of exchange. In methyloctane cracking on offretite a higher C_3/C_4 ratio is obtained for high levels of $H^{\oplus}$ exchange, since under those circumstances the reactant molecule can easily enter the gmelinite cages. If some potassium ions are present in the crystals, the C_3/C_4 ratio is reduced.

Erionite is a zeolite with intracrystalline space accessible to molecules having a critical dimension not larger than that of a normal paraffin. When using this zeolite to crack n-paraffins of chain length ranging from C_{10} to C_{23}, the product distribution is not the one predicted by the carbenium ion theory. Indeed, when cracking n-$C_{22}H_{46}$, no products in the range C_1 to C_2, C_7 to C_9, or longer than C_{12} are obtained [48]. In the case of $C_{23}H_{48}$ the principal cracked products are in the range C_{11} to C_{12}. In the case of n-decane, n-dodecane, and n-tetradecane, C_9 to C_{10} are the predominant products. These results have been attributed to a "cage effect." It is postulated that this effect increases the residence time of the molecule in the erionite when the length of the molecule approaches that of the framework cages [48,49]. The

Table 5.12   Initial Conversion of Hexadecane at 450°C

Catalyst	HZSM-5	HNu-1A	HNu-1B	HFu-1
Conversion	>99	95	60	90

Source: Ref. 51.

cage effect is also suspected to be behind the observation that n-hexane, in a mixture of C_5 to C_8 paraffins, is preferentially cracked on erionite [50].

The zeolites with intermediate pore-size openings adsorb straight-chain paraffins, isoparaffins, and monocyclic hydrocarbons at room temperature faster than they adsorb paraffins with quaternary carbon atoms. Spencer and Whittan [51] have compared the initial cracking activity for hexadecane on a series of intermediate-pore-size zeolites. Their results are presented in Table 5.12.

In the case of amorphous silica alumina, the initial conversion was 80%. The differences in activity can be explained if one takes into account that HFu-1 and HNu-1 zeolites have slightly smaller pores than HZSM-5 and that the pores of HNu-1B are smaller than those of HFu-1 [52].

From the applied point of view, the ZSM-5 zeolite is perhaps the most interesting of the medium-pore-size zeolites. Its cracking activity for a mixture of n-hexane and 2-methylpentane has been compared [53] with that of an erionite which only admits straight-chain hydrocarbons, as shown in Table 5.13.

Clearly, erionite is more selective against branched chains than ZSM-5. For the ZSM-5 zeolite, the following general trend (p. 147) in cracking rates in a mixture of C_5 to C_7 paraffin isomers is observed [54].

Table 5.13   Shape Selectivity of Erionite Versus ZSM-5

	Erionite	ZSM-5
n-Hexane	92	98
2-Methylpentane	7	56

Source: Ref. 53.

With respect to chain length:

$$n\text{-}C_7 > n\text{-}C_6 > n\text{-}C_5$$

Among the isomers:

$$n\text{-}C_7 > 2\text{-methyl-}C_6 > 3\text{-methyl-}C_6 > \text{dimethyl-}C_5 \text{ and ethyl-}C_5$$

$$n\text{-}C_6 > 2\text{-methyl-}C_5 > 3\text{-methyl-}C_5 > \text{dimethyl-}C_4$$

Therefore, the rate of conversion for each group of molecules increases with increasing molecular chain length and with molecular complexity. This indicates that in the case of ZSM-5, the diffusion rate is controlled by the shape of the molecule. Another result reported in this work is that shape selectivity decreases as the reaction temperature increases. Below 370°C, ZSM-5 cracks n-hexane 10 times faster than 3-methylpentane and 100 times faster than 2,3-dimethylbutane; at 500°C the three hexane isomers are cracked at approximately the same rate.

Drawing on these and other works, the most important characteristics of ZSM-5 as a cracking catalyst include its reported ability to produce a high-octane gasoline, the low amounts of coke formed, its ability to combine the aromatics with olefinic fragments generated in the cracking of paraffins to produce alkylaromatics, and the decrease of the paraffin cracking rate with increasing blending octane number [53,55–58] of the paraffin.

It is now obvious that medium-pore-size zeolites present some properties that make them interesting as cracking catalysts. This has opened a new and active field of research on "shape-selective" catalysis and its many important catalytic effects, which may be due to pore size, pore tortuosity, internal surface configuration, sub-Knutsen diffusion, and cavity dimensions. These must be added to the complexities already introduced by the nature, number, strength, and location of active sites.

5.2.4 Mechanism of Catalytic Cracking of Paraffins

The first event in the cracking of paraffins is the formation of the carbocation. Once the ion is formed, the cracking of a carbon-carbon bond proceeds with the formation of a smaller adsorbed carbocation and a gas-phase olefin. The cracking itself follows the β rule for carbenium ions, which is to say that the carbon-carbon bond that is cracked is in the β position with respect to the atom that contains the positive charge. The olefin formed is an α-olefin:

$$R-CH_2-CH_2-CH_2-CH_2-R' \xrightarrow{\text{cat.}}$$

$$R-CH_2-\overset{\oplus}{CH}-CH_2-CH_2-R' \longrightarrow$$

$$R-CH_2-CH = CH_2 + R' \overset{\oplus}{CH_2}$$

The newly formed carbenium ion is highly unstable and can either desorb by accepting an $H^{\ominus}$ from the active site or isomerize to an interior carbenium ion in order to stablize the charge. Although this mechanism is almost universally accepted, there remains the question of how the original carbocations are formed.
Olah et al. have shown that on cracking alkanes in superacids the initial step involves the protonation of the alkane to form a carbonium ion [59−63]. In heterogeneous catalysis, on the other hand, it has been proposed that a carbenium ion is formed initially on a Lewis site by hydride ion abstraction [64,65].

In the case of zeolite-based cracking catalysts, several hypotheses of initiation have been proposed:

1. The carbenium ion is formed by the abstraction of a hydride ion on a Lewis acid site [18,66,67].

2. The carbenium ion is formed by the abstraction of a hydride ion by a strong Brønsted acid site with the resultant formation of hydrogen as product [68−71].

3. The original ion is a penta-coordinated carbonium which is formed by the addition of a proton from a strong Brønsted site on the catalyst [5,69].

4. The carbenium ion is due to the adsorption of thermally produced olefins on Brønsted sites [69,70,72−74].

5. The carbocation is formed by the polarization of a feed molecule by strong electric fields in the pores of the zeolite [17].

Various experimental studies have been carried out to check out these suggestions. It has been reported that the OH groups present in the supercavity of an HY zeolite are consumed during the catalytic cracking of alkanes [75,76]. This suggests that protonation is taking place. Anufriev et al. [77] have observed an increase in the rate of alkane cracking when olefins are added to the feed. Perhaps during cracking the protons are being consumed by product olefins to form chain-initiating surface species.

These and many other such studies have all failed to produce an unequivocal mechanism of initiation. It would appear that the formation of carbenium ions on a Brønsted site with the consequent release of hydrogen would be the easiest hypothesis to confirm or deny. However, because much of the earlier work is confused by thermal reactions that occur simultaneously with the catalytic reactions, the presence of hydrogen in the products of catalytic cracking is not necessarily a good indicator of the validity of this hypothesis.

The initiation theory which has the highest popularity is that the thermal production of olefins by cracking in the gas phase is followed by their adsorption on Brønsted acid sites. To test this hypothesis, Weisz reports the following two experiments [78]. In the first experiment a paraffin is passed over a cracking catalyst; in the second the paraffin is first passed over a hydrogenation-dehydrogenation catalyst and then over a second bed containing the same cracking catalyst that was used in the first experiment. It was found that the rate of reaction was less in the second experiment than in the first. This evidence has convinced many authors that the formation of the initial carbenium ions on the cracking catalyst proceeds from thermally produced olefins which, it is argued, are saturated on the hydrogenation catalyst of the second experiment.

Pines and Wackher [79] have observed that when n-butane reacts in the presence of $AlCl_3$ + HCl at 100°C, conversion increases from 12% to 27% when the initial concentration of admixed butenes increases from 0.015 mol % to 0.3 mol %. However, others [80] have shown that in the case of the isomerization of n-butane on superacid polymers, the presence of 1-butene had little effect on the isomerization rate. Thus the role of thermally produced olefins remains uncertain at this time. Even if their role in initiation is accepted, there seems to be more to initiation than just this one possibility.

Corma et al. [10] have shown that the initial ratio of paraffin to olefin in the products of n-heptane cracking on REHY is significantly greater than the expected value of 1. This ratio decreases with conversion up to about 5% and then begins to increase. Such behavior cannot be explained solely by the β cracking of carbenium ions. Clearly, a source of hydrogen that exists on the fresh catalyst is used up in the early stages of the reaction and is then replaced at higher conversions. We do not have to look far for the source of hydrogen at high conversions. It is the coke which, on forming, begins to supply hydrogen to the product. The initial hydrogen must come from the catalyst itself and more than likely from OH groups which are present on the surface.

Using molecular orbital calculations, the authors have simulated the interaction of a proton from the OH with n-heptane. They report [10] that protolytic cracking is energetically favorable by 20 kcal/mol over β scission via carbenium ions of n-heptane and on a par energetically with the cracking of branched heptanes. Thus initiation by protolytic cracking is a distinct possibility here and leads to the idea that the mechanism shown in Figure 5.4 may

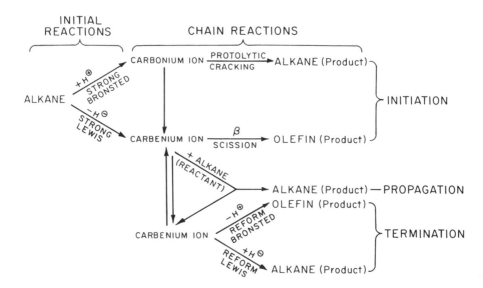

Figure 5.4   Processes involved in alkane cracking on zeolites.

be operating in this, and probably all, cracking reactions on solid catalysts.

In Figure 5.4 we see that the initial alkane-to-olefin ratio will depend on the initial Brønsted/Lewis ratio on the catalyst. The first cracking event will then produce a certain alkane-to-olefin ratio according to the ratio of reactions 1 and 2. This leaves carbenium ions on what used to be Lewis and Brønsted sites. If these were to desorb, the ratio of alkane to olefin would be restored to 1. However, if a chain of reactions is initiated on at least some of the sites, an initial high yield of alkanes will be produced. This process can also be envisioned as a formation of paraffins by the dehydrogenation of surface coke.

5.2.5   Rate-Controlling Step in Paraffin Cracking

It is widely believed [68] that the formation of the carbocation is the controlling step in the cracking of paraffins and other hydrocarbons. This belief is largely based on the fact that it is easier to form a carbocation on an olefin than on a paraffin. Since olefins crack faster, it is inferred that the faster rate-controlling step (i.e., the formation of carbenium ions in olefins) results in the faster overall reaction. This reasoning is based on two unstated assumptions: that the carbocations are formed on the same sites for both olefins and paraffins, and that the nature of the carbocation formed is the same in olefins and paraffins. Neither assumption has been proven, nor is there reason to believe that either is true. Indeed, it seems that in the case of solid acids the active sites for the cracking of paraffins are both Lewis acid sites and strong Brønsted sites, while Brønsted acid sites alone are active in the cracking of olefins. Furthermore, the carbocations formed in these two cases do not have to be the same, if only because their counterions are not the same. The picture is further complicated by the fact that there is evidence which points to the evolution of the carbocation rather than to its formation as the rate-controlling step [8,11].

The problem of the rate-controlling step has been tackled by studying the mechanism of heptane cracking [3] both experimentally and by molecular orbital calculations. It was shown experimentally that the activation energies for the cracking of n-heptane into $C_2 + C_5$ and $C_4 + C_3$ differ by 11 kcal/mol. On the other hand, the differences in the calculated energy of formation of the two carbocations which lead to $C_2 + C_5$ and $C_4 + C_3$, respectively, is only about 1 kcal/mol. If the controlling step is the formation of the parent carbocation, the difference in activation energies for the two possible splits would be closer to 1 kcal/mol than the observed 11 kcal/mol. The conclusion in this case is that the controlling step is not the formation of the parent carbocation but its subsequent evolution on the surface. Others [64,65] have shown that in the case of the isomerization of pentanes on a solid superacid, the carbocation rearrangement is a slow step, whereas the formation of the carbocation, by extraction of an $H^{\ominus}$ from the reactant, is very fast. At the same time the authors suggest that in the case of metal halide catalysts, abstraction of $H^{\ominus}$ from alkanes may very well be the slow step in the reaction.

Despite the general agreement about the products of reaction and the generalities of the influence of molecular structure on the rate of reaction, there is no general agreement on the fundamental question of what sites are responsible for the initial formation of the carbocations in alkane cracking, or which is the rate-controlling step in the cracking of paraffins.

5.3   CATALYTIC CRACKING OF OLEFINS

In the cracking of olefins it is generally agreed that the first event is the formation of a carbenium ion by the addition of a proton from a Brønsted site to the double bond of the olefin. Once formed, the carbenium ion cracks following the β rule. This event produces a smaller olefin and a smaller adsorbed primary carbenium ion. The resultant primary carbenium ion will either rearrange into a secondary carbenium ion or desorb:

$$R-CH_2-CH = CH-CH_2-CH_2-R' + H^{\oplus} \longrightarrow$$

$$R-CH_2-\overset{\oplus}{CH}-CH_2-CH_2-CH_2-R' \longrightarrow$$

$$R-CH_2-CH = CH_2 + R'-CH_2-\overset{\oplus}{CH}_2 \longrightarrow R'-CH = CH_2 + H^{\oplus}$$

As mentioned before, olefins are observed to crack much faster than paraffins of corresponding chain length and configuration [66].

In the case of short-chain olefins such as C_2-C_5, cracking is observed to be less important than reactions involving hydrogen transfer and polymerization yielding paraffins and coke. For example, ethylene at 400°C on a silica-alumina catalyst [81] forms some 2% C_3 and C_4 hydrocarbons and 6% ethane while depositing some 9.6% coke on the catalyst. The same reaction with n-butene yields 7% C_3, 7% isobutene, and 8% isobutane, while 20% of the feed is converted to liquid products and 6% to coke.

It is well recognized that olefins are an important source of carbon formation on cracking catalysts. In particular, the short-chain olefins appear to lead to coke as one of the major products of reaction. Since all cracking reactions yield olefins among the products and since all cracking in due course leads to short-chain olefins among the products, it may well be that it is these short-chain olefins that are responsible for much of the coke formation in catalytic cracking. For instance, it is reported [82] that up to 27% of butadiene feed is converted to coke when it is cracked over a series of sodium-ammonium Y catalysts calcined at various temperatures. Others [83,84] have shown that ethylene adsorbs very strongly on Brønsted acid sites on HY zeolites even at low temperatures. When the temperature is raised, proton transfer and isotopic exchange occurs in the adsorbed ethylene, testifying to the activity of the surface species.

It is surprising how few papers there are on the cracking of olefins compared to the cracking of paraffins. This interesting topic needs to be pursued with vigor and promises to hold the key to the details of catalytic coke formation.

5.3.1 Cracking of C_5, C_6, and C_7 Olefins

Several authors report that the cracking of n-pentene proceeds at 400°C exclusively by a process of dimerization followed by cracking of the C_{10} [13–15,85,86]. Corrected initial selectivities observed in the cracking of normal C_5, C_6, and C_7 olefins on ZSM-5 are shown in Table 5.14, together with the selectivities for skeletal rearrangement (SR). There we see that C_5 cracks very slowly compared with its rate of isomerization and that longer molecular chains show an increased proportion of cracked products. Furthermore, C_5 cracks exclusively by dimerization followed by cracking, as evidenced by the initial products, which are solely C_3 and C_4, and by the ratio of their initial selectivities, which is about 2. C_7 in turn cracks almost exclusively by a monomolecular process, as evidenced by the fact that its C_3/C_4 ratio is about 1.

Hexene is the transition case at this temperature and cracks by both mono- and bimolecular processes. In all these cases there was no evidence of significant oligomerization or aromatization.

Table 5.14   Corrected Initial Selectivities for
Cracking of n-Alkenes on ZSM-5 at 405°C

	1-Pentene	1-Hexene	1-Heptene
C_3	0.0173	0.0201	0.0468
C_4	0.0095	0.0175	0.0509
C_5	(SR)	0.0130	0.0097
C_6	—	(SR)	—
C_7	—	0.0025	(SR)
C_8	—	0.00098	—
C_9	—	—	—
	0.0268	0.05408	0.1074
SR	0.9710	0.9444	0.8974
	0.9978	0.9985	1.004

Source: Ref. 85.

A more detailed examination of the data shows that the ratio of mono- to bimolecular rates is 0, $\sim$0.2, and $\sim$7 for C_5, C_6, and C_7 olefins, respectively. This compares with $\sim$1.5 for saturated n-C_7 at the same temperature [2]. It therefore seems that olefins crack more directly than do the corresponding paraffins. At the same time, C_7 olefin isomerization constitutes >89% of the initial reaction on ZSM-5, whereas n-heptene shows only 17% isomerization under initial conditions on HY. Further work will reveal whether these differences are due to differences in the two catalysts or to the carbocations formed on the two types of molecules.

The total absence of C_2 as an initial product in the cracking of olefins on ZSM-5 indicates that a significant difference in stability exists between the $C_2^{\oplus}$ and $C_3^{\oplus}$ ions. Furthermore, hydrogen production was zero under initial conditions, indicating that abstraction of a hydride ion by strong Brønsted sites does not take place.

Figure 5.5 shows the network of isomerization reactions that occur in n-heptene cracking on ZSM-5 together with the initial selectivities for the various products. One observes that the

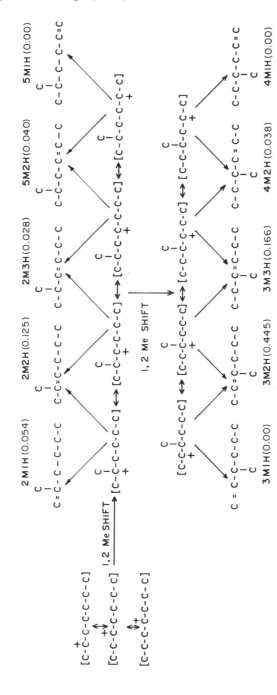

Figure 5.5 Network of skeletal rearrangements in *n*-heptene cracking. Quantities in brackets indicate initial selectivities. (From Ref. 85.)

Table 5.15   Initial Product Selectivities in 1-Hexene
Isomerization at 200°C

Product	Initial selectivity			Reaction type
	HY (200°C)	ZSM-5 (200°C)	ZSM-5 (280°C)	
t-2H	0.500	0.620	0.638	DBS
c-2H	0.415	0.335	0.267	DBS
(c + t)-3H	0.062	0.030	0.090	DBS
1P	0.00032	0.00080	0.0022	(P + C)
1B	0.00070	—	—	(P + C)
iB	0.0051	—	0.0012	(P + C)
Pre	0.00060	—	0.00070	(P + C)
c-2B	—	—	0.00050	(P + C)
t-2B	—	—	0.0010	(P + C)
(c + t)-2P	—	—	0.00040	(P + C)
2MBa	0.00033	—	—	HT
iBa	0.00028	—	0.00038	HT
Pra	0.0060	—	—	HT
Ea	0.0014	—	—	HT
Coke	0.0055	0.0340	0.0050	P, C, CZ
Totals				
DBS	0.977	0.985	0.995	
HT	0.00801	0.0	0.00038	
P + C	0.00672	0.00080	0.0050	
Coke	0.0055	0.0340	0.0050	

Source: Ref. 88.

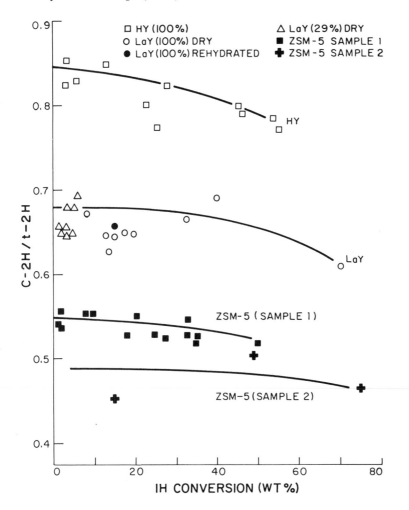

Figure 5.6   Cis/trans ratios in hexene isomerization on various catalysts in a variety of states of hydration.   (From Ref. 88.)

dominant initial products are the ones that arise from tertiary
carbenium ion precursors. Furthermore, it seems that the tertiary
ion on the third carbon is more stable than that on the second
carbon of the chain.

5.3.2   Isomerization of *n*-Hexene-1

The isomerizations of normal olefins can be studied in isolation
from cracking at temperatures of about 200°C [87,88]. Studies
of *n*-hexene-1 isomerization have shown that the reaction proceeds directly to hexene-2 and hexene-3. Table 5.15 shows
initial selectivities on HY and ZSM-5. Studies on many other
catalysts have shown that the cis/trans hexene-2 ratio is a
property of the catalyst under study, as shown in Figure 5.6.
It appears that the initial cis/trans ratio is structure sensitive
rather than being dependent on the activity of the catalyst.
When LaY is dehydrated its activity for the isomerization reaction
drops dramatically without causing the ratio to depart from the
LaY line shown in Figure 5.6. We are therefore encouraged to
think that steric properties in the vicinity of the active sites
themselves are responsible for governing the preferred configuration of the product hexene-2.

   The fact that large losses in activity occur when LaY is dried
under conditions where cracking would normally take place indicates that weakly bound water is responsible for much of the
double-bond shifting activity in hydrated catalysts at 200°C.
Since such water is known to generate weak sites, one can assume that this type of isomerization proceeds on weak sites.

5.4   CATALYTIC CRACKING OF CYCLOPARAFFINS

The primary product of the cracking of cycloparaffins is the corresponding olefinic chain isomer. This can be seen most readily
in the case of small cycloparaffins such as cyclo C_3's and C_4's,
in which the primary product can easily be isolated. For instance,
Basset and Habgood [89] found the primary product of the cracking of cyclopropane to be propylene. Because of its great simplicity, this is a very attractive, although underutilized test reaction for cracking catalysts [90—96]. Tam et al. [97] have suggested the following mechanism to explain the results obtained by
various authors:

$$CH_2 \diagdown CH_2 \diagup CH_2 \underset{\underset{Z^{\ominus}}{\overset{H^{\oplus}}{\diagup}}{\diagdown}CH_2 \longrightarrow CH_3 - CH = CH_2 + \left[H^{\oplus}Z^{\ominus}\right]$$

$$\searrow CH_2 = CH_2 + \left[CH_3^{\oplus}Z^{\ominus}\right]$$

$$\left[CH_3^{\oplus}Z^{\ominus}\right] + C_3H_6 \longrightarrow \underset{H_3C}{\overset{H_3C}{\diagdown}}C = CH_2 + \left[H^{\oplus}Z^{\ominus}\right]$$

Kiricsi et al. [98] report that the cracking of cyclopropane over
NaHY leads to propylene as the only primary product; isobutane,
2-methylbutane, and 2-methylpentane are secondary products of
the reaction. No C_1 or C_2 hydrocarbons were found, and C_5,
C_6, and higher ring systems react at a slower rate than the corresponding paraffins or primary olefins. The activation energies
for cracking of n-hexane and cyclohexane are nevertheless very
close [6], as are the activation energies for n-heptane and methylcyclohexane cracking [4,99]. However, the distribution of products in the paired cases is markedly different. On zeolite catalysts, the cracking of naphthenes gives a large amount of isomerization, and significant amounts of aromatic products are obtained
[99,100]. This same type of distribution of products was found by
and large in alkylcyclohexanes, dialkylcyclohexanes, dicyclohexyl-
decaline, cyclohexane, and methylcyclopentane [81]. It should be
noted that the amount of hydrogen found in the products of such
cracking is greater than in the case of the corresponding paraffins.
Greensfelder and Voge [11] found 27% hydrogen plus C_1 to C_4
products in the cracking of methylcyclohexane, while only 6.5%
were found in the cracking of n-heptane. They were led to the
conclusion that dehydrogenation must be taking place in this case.
This is a surprising conclusion in view of the fact that in all
these cases the catalysts are purely acid catalysts and do not contain any metal hydrogenation-dehydrogenation functions. The
point was studied by various authors [99,100] who found that such
catalysts can be active in dehydrogenation of cycloaromatics, although the mechanism of the reaction is not clear. As we see in
the mechanism shown in Figure 5.7, the carbocation of a cyclo-
paraffin can rapidly be converted to a mixture of isomeric ions,
which may lead to cracking, dehydrogenation, or ring contraction.

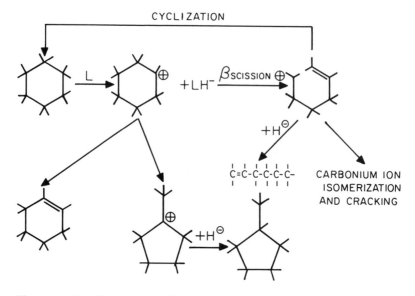

Figure 5.7 Network of the reactions involved in the cracking of cyclohexane.

The observed hydrogen-donating ability of cycloparaffins in cracking is well illustrated by the high ratio of paraffin to olefin obtained in the products when a mixture of cycloalkanes and olefins or paraffins is cracked [11,101]. The recognized ability of cycloparaffins to donate hydrogen has even been used to suppress coke formation by the addition of tetraline to cracking mixtures [102].

This branch of cracking research may yet reveal a new type of active site whose presence on cracking catalysts leads to dehydrogenation and coke formation.

5.5 CATALYTIC CRACKING OF ALKYL AROMATICS

In the cracking of alkyl aromatics the benzene ring remains unaltered, while the side chain in all cases except that of toluene is cracked off, yielding an olefin. The effect of chain length, chain branching, and the position of associated alkyl groups is illustrated in Table 5.16.

Table 5.16   Effect of Branching on Cracking
Activation Energy

Hydrocarbon	Activation energy (kcal/mol)
$C_6H_5-C_2H_5$	50
$C_6H_5-CH_2-CH_2-CH_2-CH_2$	34
$C_6H_5-\underset{\underset{CH_3}{\mid}}{CH}-CH_2-CH_3$	19
$C_6H_5-\underset{\underset{CH}{\mid}}{CH}-CH3$	17.5

Source: Ref. 103.

In general, the rate of cracking of the side chain increases as one goes from primary to secondary to tertiary attachment of the chain to the ring, while in the case of the same type of attachment, the rate increases with increasing size of the attached chain.

In the case of toluene, the dominant reaction is the disproportionation to benzene and xylenes [104], not cracking to methane. In the case of polymethylbenzenes, the dominant reaction is the isomerization and disproportionation of the parent molecules. For example, dimethylbenzenes in the presence of acid catalysts isomerize and disproportionate according to the following mechanism [105]:

o-xylene $\longrightarrow$ M-xylene $\longrightarrow$ p-xylene

toluene + trimethylbenzene

In the case of ethylbenzene, it has been shown that transalkylation accompanies the cracking reaction at temperatures above 200°C. At lower temperatures only transalkylation is observed [105,106]. Thus it is reasonable to suppose that in all alkylaromatics there is competition among cracking, transalkylation, and the various isomerization reactions.

5.5.1 Cracking of Cumene

Of all the alkyl aromatics, the one that has been studied most extensively is cumene or isopropylbenzene. This reaction has been long utilized as a standard test reaction in the study and evaluation of cracking catalysts. A review by Corma and Wojciechowski [107] gives details of the state of knowledge of cumene cracking.

The use of test reactions has been common in the field of catalytic cracking for many years. The need for such reactions arises from the fact that the feed used commercially is far too difficult to analyze and its reactions too complex and too numerous to understand in detail. Various reactions have been proposed and used in determining the properties of cracking catalysts in the laboratory. The essential characteristics of a test reaction for the proper evaluation of a catalyst are as follows:

1. The reaction must proceed by a mechanism that is typical of the cracking of some of the components of a petroleum feedstock.

2. The reaction must yield a minimum of products so that a complete analysis and a quantitative treatment of the mechanisms involved can be undertaken.

The cracking of cumene complies with these requirements in that it represents the reactions of alkyl aromatics, proceeds via a carbocation, and yields few products in significant quantities, benzene and propylene being the dominant species.

In most early studies, for instance that of Prater and Lago [108], benzene and propylene were in fact the only products observed. However, various authors soon began reporting the presence of other products, such as diisopropylbenzene and mixed alkanes [109,110]. Richardson [111] identified a great variety of products in the cracking of cumene on faujasite exchanged with alkali earth cations, including benzene, toluene, α-methylstyrene, propylene, acetylene, ethylene, and methane. Rabo and Poutsma report the presence of propane; propylene; C_4, C_5, and C_6 aliphatics; benzene; toluene; and ethylbenzene. More recent studies [112,113] have identified the products shown in Table 5.17.

Table 5.17   Major Products Observed
During the Cracking of Cumene on
LaY and HY Zeolite Catalysts

Liquids	Gases
Benzene	Methane
Toluene	Ethane
Ethylbenzene	Ethylene
o-Ethyl toluene	Propane
n-Propyl benzene	Propylene
Cymenes	Butane
m-Diisopropylbenzene	Butene
p-Diisopropylbenzene	i-Butane
	i-Butene

Source: Ref. 113.

Thorough analysis of the reaction schemes involved in the production of the products noted above leads to the conclusion that
on the various catalysts studied there are two groups of reactions,
each apparently occurring on a different type of site [114].   The
overall network of reactions involved in the cracking of cumene is
shown in Figure 5.8.   At high temperatures this bewildering array
of reactions reduces to the simple delta mechanism shown in Figure 5.9, which accounts for the cracking of cumene per se.

A detailed study of the kinetics of this reaction has shown that
the rate-controlling step is the decomposition of the surface carbocation into gas-phase benzene and an adsorbed propylcarbenium
ion [115].   Thus, in this case at least, the formation of the
carbonium ion itself is not the rate-determining step.

The study of cumene cracking has resulted in the development
of powerful techniques for mapping catalyst site activity.   Since
these techniques are broadly applicable and of broad potential
utility, they deserve closer examination.

To begin, selectivities of all the products are calculated from
the OPEs and used to derive a reaction network.   Figure 5.10
shows selectivity curves for the various products observed in
cumene cracking and illustrates the large and seemingly confused
differences between the cracking behavior of LaY and HY.

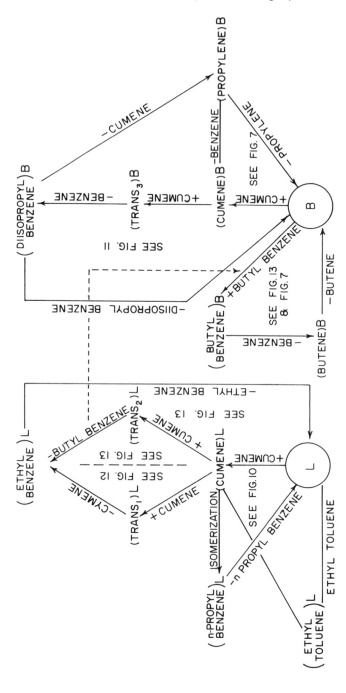

Figure 5.8   Network of primary processes taking place during the cracking of cumene.   B and L denote Brønsted and Lewis sites.   (From Ref. 107.)

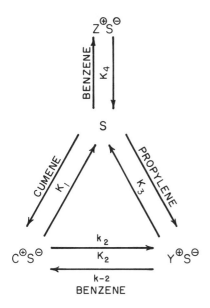

Figure 5.9 Delta mechanism of cumene cracking. (From Ref. 115.)

From the reaction network it can be seen that to calculate the corrected selectivity for benzene production by dealkylation alone, one must subtract from the total benzene yield the benzene produced by disproportionation to diisopropylbenzenes and in the production of C_4. Using this correction, benzene yields at all conditions are calculated and fitted by a mechanistic model to yield an initial rate of benzene production by dealkylation alone. By dividing this rate by the corrected initial selectivity for benzene shown in Table 5.18, one obtains the initial rate of total cumene conversion. Such total conversion rates are shown for various catalysts in Table 5.19.

How well the calculated dealkylation rates fit the observed conversion of cumene is illustrated in Figure 5.11. There we see that the fit is very good over a wide range of conversions and catalyst activities, as reflected by the time-on-stream axis.

Using the total rates given in Table 5.19 and the selectivities for the minor products shown in Table 5.18, one can calculate the rate constants for the production for all primary products. Note that the bimolecular rate constants obtained in this way have to be divided by the initial concentration of cumene to avoid having

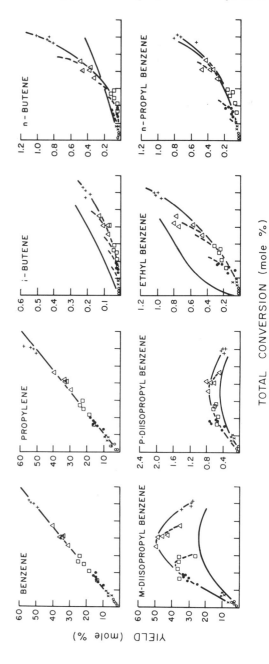

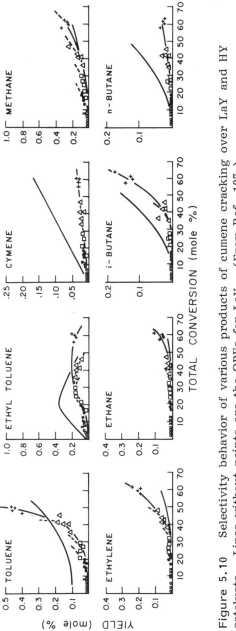

Figure 5.10 Selectivity behavior of various products of cumene cracking over LaY and HY catalysts. Lines without points are the OPEs for LaY. (From Ref. 107.)

Table 5.18 Initial Selectivities for Primary Products in the Catalytic Cracking of Cumene on HY and LaY Zeolite Catalysts

Product	Temperature (°C)					
	360 Initial selectivity		430 Initial selectivity		500 Initial selectivity	
	HY	LaY	HY	LaY	HY	LaY
Benzene[a]	0.61	0.65	0.84	0.87	0.94	0.94
Propylene[a]	0.63	0.64	0.84	0.87	0.92	0.93
n-Propyl benzene[a]	0.0016	0.0030	0.0015	0.0031	0.0014	0.0038
Cymene[b]	0.0016	0.012	0.00048	0.0033	0.00015	0.0010
m-Diisopropylbenzene[b]	0.128	0.080	0.050	0.036	0.016	0.014
p-Diisopropylbenzene[b]	0.068	0.064	0.027	0.020	0.0078	0.0068
n-Butene[b]	0.0022	0.0095	0.0016	0.0040	0.00090	0.0037
i-Butene[b]	0.0020	0.0070	0.00080	0.0048	0.00025	0.0018

[a]Monomolecular.
[b]Bimolecular.
Source: Ref. 114.

Table 5.19   Rates of Cumene Conversion on 70/80 Mesh Catalyst

Catalyst	Rate constant $\times 10^4$ (mol$\cdot$g^{-1} s^{-1})		
	360°C	430°C	500°C
HY	60.66	315.48	1,521.28
LaY	26.15	121.8	505.32
Si/Al (25% Al)	—	—	29.52
Si/Al (13% Al)	—	—	7.00

this term distort activation energies. In this way the rate constants shown in Table 5.20 were obtained. By plotting such rate constants on an Arrhenius plot, one can obtain activation energies for all the primary processes. Figure 5.12 shows that for dealkylation on HY and LaY the activation energy is the same, for all practical purposes. This is true for all the products, although the values reported in Table 5.21 were obtained by statistical fitting to the three temperatures available.

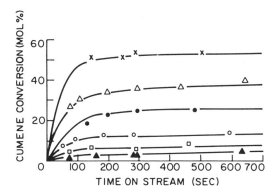

Figure 5.11   Least-squares fit for cumene dealkylation data on 70/80 mesh catalyst at 430°C. (From Ref. 113.)

Table 5.20 Initial Kinetic Rate Constants for Primary Reactions in the Catalytic Cracking of Cumene on 70/80 Mesh HY and LaY Zeolite

Product	360°C $k_0[S_0]\left(\dfrac{mol^{2-n}}{g^{2-n}\cdot s}\right)$		430°C $k_0[S_0]\left(\dfrac{mol^{2-n}}{g^{2-n}\cdot s}\right)$		500°C $k_0[S_0]\left(\dfrac{mol^{2-n}}{g^{2-n}\cdot s}\right)$	
	HY	LaY	HY	LaY	HY	LaY
Benzene[a]	37.0×10^{-4}	17.0×10^{-4}	265.0×10^{-4}	106.0×10^{-4}	1430.0×10^{-4}	475.0×10^{-4}
n-Propyl-benzene[a]	0.09×10^{-4}	0.08×10^{-4}	0.47×10^{-4}	0.50×10^{-4}	2.10×10^{-4}	1.93×10^{-4}
Cymene[b]	0.20×10^{-9}	1.65×10^{-9}	0.26×10^{-9}	0.68×10^{-9}	0.40×10^{-9}	0.84×10^{-9}
m-Diisopropyl-benzene[b]	15.80×10^{-9}	4.25×10^{-9}	27.40×10^{-9}	7.74×10^{-9}	40.2×10^{-9}	11.70×10^{-9}
p-Diisopropyl-benzene[b]	8.53×10^{-9}	3.40×10^{-9}	14.80×10^{-9}	4.33×10^{-9}	19.6×10^{-9}	5.70×10^{-9}
n-Butene[b]	0.26×10^{-9}	0.39×10^{-9}	0.87×10^{-9}	0.99×10^{-9}	2.30×10^{-9}	3.10×10^{-9}
i-Butene[b]	0.24×10^{-9}	0.51×10^{-9}	0.43×10^{-9}	0.85×10^{-9}	0.60×10^{-9}	1.51×10^{-9}

[a]Monomolecular.
[b]Bimolecular.
n is the order of reaction.
Source: Ref. 114.

Table 5.21 Activation Energy for Primary Reactions in the
Catalytic Cracking of Cumene on HY and LaY Zeolite Catalysts

Product	Molecularity of the reaction	Activation energy (kJ/mol) HY 70/80 Mesh	LaY 70/80 Mesh	LaY 100/140 Mesh
Benzene	1	106.7	97.5	97.5
m-Diisopropyl-benzene	2	27.2	29.7	26.8
p-Diisopropyl-benzene	2	24.3	15.1	18.8
Cymene	2	20.1	8.4	22.2
n-Propylbenzene	1	95.4	93.3	90.0
n-Butene	2	63.6	60.2	61.9
i-Butene	2	27.2	31.8	34.7

Source: Ref. 114.

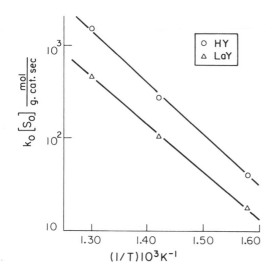

Figure 5.12 Activation energy plot for cumene cracking on HY
and LaY. (From Ref. 107.)

171

The catalytic rate constants calculated in this way are insep-
arable from the active-site concentration and have the structure

$$k_0 = Ae^{-E/RT}[S_0] =$$

(frequency factor) $(e^{-E/RT})$(initial site concentration)

(5.1)

The constants k_0, if plotted on an Arrhenius plot, will give
the activation energy of the reaction provided that A and $[S_0]$
are not functions of temperature. We need not worry about A,
but steps have to be taken to assure the constancy of $[S_0]$. This
is most conveniently done by regenerating the catalyst at the
same temperature after each run and making that temperature
somewhat higher than the highest reaction temperature to be investigated. This procedure makes sure that the initial Brønsted/
Lewis site ratio is the same in each run and that the total $[S_0]$ is
the same. This is an important point, because in principle the
technique is capable of tracing changes in these quantities with
changes in regeneration procedure.

Once these points have been accommodated, there is good reason to believe that the differences between the catalysts are due
to differences in the frequency factors or in active-site concentrations. Since there is no reason to suspect that for a given reaction the frequency factor will change without an associated
change in the activation energy, one can safely conclude that the
differences are due to differences in active-site concentrations.
Furthermore, since the rate constants have the same A and E on
each catalyst, we conclude that the sites responsible for dealkylation are the same in each catalyst. Proceeding in this manner with
the other primary products, the activation energies for the production of each of them are calculated and are shown in Table
5.21.

As mentioned before, all activation energies on all the catalysts
shown in Table 5.21 are the same for any given product. This
leads to the conclusion that on each catalyst a given reaction proceeds on exactly the same type of site. There is no other way of
demonstrating this fact and little doubt that this conclusion is
warranted by the data.

The next question that needs to be examined is whether all the reactions compete for the same sites or whether they proceed on their
own specific sites. The fact that they all have different activation

Table 5.22 Ratios Between Kinetic Rate Constants for HY and LaY on 70/80 Mesh

	Temperature[a] (°C)		
	360	430	500
Product	HY/Lay	HY/LaY	HY/Lay
Benzene	2.2	2.5	3.0
m-Diisopropyl-benzene	3.7	3.5	3.4
p-Diisopropyl-benzene	2.5	3.4	3.4
Average	2.8 ± 0.65	3.1 ± 0.42	3.3 ± 0.22
Cymene	0.34	0.38	0.48
n-Propylbenzene	0.87	0.94	1.1
n-Butene	0.67	0.88	0.74
i-Butene	0.47	0.51	0.40
Average	0.59 ± 0.20	0.68 ± 0.29	0.68 ± 0.28

[a] Average for the three temperatures plus or minus the standard deviation: $k_{HB}/k_{LaB} = 3.1 \pm 0.49$; $k_{HL}/k_{LaL} = 0.65 \pm 0.26$.
Source: Ref. 114.

energies is simply a reflection of the energetics of each process and tells us nothing of site quality. To test whether all reactions proceed on the same sites, one must adopt a standard catalyst, say LaY, and then for each of the other catalysts, take ratios of the rate of each primary reaction, at each temperature, to the corresponding reaction rate on LaY. Table 5.22 shows one such result. In taking these ratios from the actual experimental points, one avoids distortions that may be caused by taking ratios of $A[S_0]$ obtained by the best fit to an Arrhenius plot.

Table 5.22 clearly shows that not all the ratios are the same. However, taking into account the experimental scatter, it would be safe to say that at least two groups of ratios appear: those whose values are greater than 1 and those whose values are less

Table 5.23   Relative Concentrations of Active Sites on Various Catalysts

Units relative to LaY	Site type	HY[a]	LaY[a]	82% LaY[b]	SA-25[c]	SA-13[c]	36% LaY[b]
Active sites per gram of catalyst	B	310	100	45	4.51	1.02	0.42
	L	65	100	—	6.90	1.66	0.22
Active sites per square meter of catalyst surface	B	310	100	45	9.08	2.07	0.42
	L	65	100	—	13.90	3.37	0.22

[a]Exchanged to approximately zero.
[b]Percent indicates exchange of lanthanum for sodium.
[c]SA is amorphous silica alumina. Number indicates percent of alumina.

Source: Ref. 140.

than 1. Taking only these two groups and averaging the values
as shown in Table 5.23, one finds that there are about three
times as many sites that produce benzene and diisopropylbenzenes
on HY and only half as many of those that produce the other
products. Using the same arguments, Table 5.23 was constructed
for various catalysts studied to date.

Details of the procedures described above are available in the
literature [107,112—115] but little use has yet been made of them.
As familiarity with the methods described in Sections 5.2, 5.3,
and 5.4 grows, much new and fundamental information on the
nature of the cracking process will become available.

5.6   OTHER REACTIONS THAT TAKE PLACE DURING CATALYTIC CRACKING

The preceding description of various hydrocarbon cracking reactions makes it amply clear that none of them involves cracking
alone. In all such reactions both the reactants and products
undergo a variety of reactions other than β scission. These reactions are all of major significance in their contribution to the commercial value of the gasoline produced by catalytic cracking and
all deserve to be studied in detail. In most cases it is not possible to study them in isolation and hence the various studies of
cracking reactions also provide a valuable source of information on
the accompanying reactions. As the techniques of initial rate
studies become more widely used, one can expect a large increase
in the somewhat limited information presently available on these
reactions.

5.6.1   Isomerization

As outlined in the discussion of carbocation reactions in the liquid
phase and in Section 5.3.2, isomerization is a relatively fast and
facile reaction. Thus it is not surprising that during the catalytic
cracking of hydrocarbons, the olefins formed are isomerized to
give the corresponding branched-chain isomers. The process of
skeletal isomerization leads to the formation of tertiary carbons,
which in turn leads to the enhancement of the cracking of the
isomerized products. As is the case in cumene cracking, it appears that the rate of isomerization is governed not by the formation of the carbocation, but by the subsequent rearrangement of
the ion on the surface [64,65].

5.6.2 Alkylation

This is the reverse process to catalytic cracking. At reaction temperatures below 400°C this process is dominant over the cracking reaction for most feeds. In fact, there is always an equilibrium between cracking and alkylation in the sense of polymerization–depolymerization. At high temperatures the equilibrium favors depolymerization and short-chain molecular products, whereas at low temperatures, polymerization, long-chain products, and catalytic coke formation are favored. In the cracking of short-chain olefins, the formation of higher-molecular-weight products is a consequence of alkylation reactions of this type. However, in the cracking of large molecular species, as for instance in the cracking of gas oil, it has been found that no polymeric products appear in the liquid products of the reaction [116].

5.6.3 Disproportionation

Condon and others [117,118] have proposed a mechanism of acid-catalyzed paraffin disproportionation involving the formation of a carbon-carbon bond between a carbenium ion and an olefin followed by the rearrangement of the resulting alkylate and the β scission of the large molecule.

$$CH_3 - CH_2 - CH_2 - CH_3 \xrightarrow{-H^{\ominus}} CH_3 - CH_2 - \overset{\oplus}{C}H - CH_3$$

$$\xrightarrow{-H^{\oplus}} CH_3 - CH_2 - CH = CH_2 \xrightarrow{CH_3 - CH_2 - \overset{\oplus}{C}H - CH_3}$$

$$CH_3 - \overset{\oplus}{C}H - CH - \underset{\underset{CH_3}{|}}{\overset{\overset{CH_3}{|}}{C}H} - CH_2 - CH_3 \longrightarrow CH_3 - CH_2 - \overset{\oplus}{\underset{\underset{CH_3}{|}}{C}} - \overset{\overset{CH_3}{|}}{C}H - CH_2 - CH_3$$

$$\longrightarrow CH_3 - CH_2 - \overset{\overset{CH_3}{|}}{\underset{\underset{CH_3}{|}}{C}} - \overset{\oplus}{C}H - CH_2 - CH_3 \longrightarrow CH_3 - CH_2 - \overset{\overset{CH_3}{|}}{\underset{\underset{CH_3}{|}}{C}} - CH_2 - \overset{\oplus}{C}H - CH_3$$

$$\longrightarrow CH_2 = CH - CH_3 + CH_3 - \overset{\oplus}{\underset{\underset{CH_3}{|}}{C}} - CH_2 - CH_3$$

Such reactions can also lead to the formation of higher hydrocarbons, as they do in the commercial process of alkylate gasoline
production. In practice, the larger molecules fall in the range
C_5 to C_{10} because of stability considerations.

The clearest examples of disproportionation occur in the exchange of alkyl groups between alkyl aromatic species, as in the
formation of diisopropylbenzene from cumene. In that case the
reaction has been shown to be primary and thus proceeds directly
by the disproportionation of two cumene molecules, resulting in
diisopropylbenzene and benzene as products [112].

5.6.4   Cyclization

Cyclization is the reverse of a well-established cracking reaction,
the cracking of cycloparaffins. The formation of cyclic hydrocarbons is thought to be due to olefins in the reacting mixture
and can lead, via sequential dehydrogenation reactions, to the
formation of aromatics as shown below or by a similar mechanism
involving allylic cations [119].

The reaction does not go selectively on ordinary catalysts because of the steric requirements. It is known that *n*-hexene-1
and *n*-heptene-1 show very little tendency to cyclize on HY or

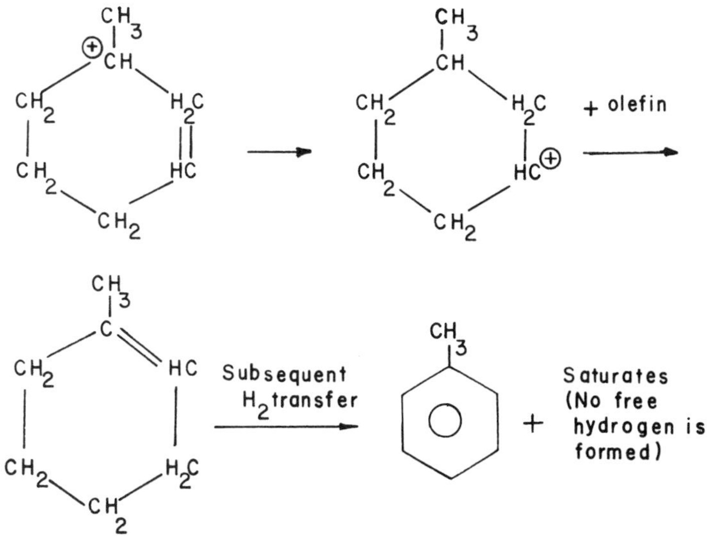

ZSM-5 [86—88], although there are reports that certain compositions of ZSM-5 do lead to aromatics from light olefins [120]. This
reaction is in all probability truly shape selective in the sense of
requiring very specific site densities and geometries.

An interesting example of successful cyclization is that of the
methanol-to-gasoline process which is carried out on   ZSM-5.
This suggests that if an appropriate feed is used, the necessary
geometry for cyclization and aromatics production is available on
this catalyst.

5.6.5   Hydrogen Transfer

The hydrogen-transfer reaction takes place between carbocations
and certain hydrocarbons. The rate of the reaction depends on
the structure of the hydrocarbon molecule that reacts with the
carbocation. Given the facility with which carbocations rearrange,
the velocity of this reaction is not dependent on the original
carbocation. The reaction is often thought to take part in a
chain transfer of charges in catalytic cracking and hence takes
part in chain propagation.

$$\overset{\oplus}{R-CH_2} + H-R' \longrightarrow R-CH_3 + R'\overset{\oplus}{}$$

It has been shown that the best hydride donors are those compounds that give rise to resonance-stabilized carbenium ions by
losing hydride ions [121]. For instance, napthenes are particu-
larly active since they are able to give up hydrogen on their way
to becoming aromatics. Voge et al. [122], cracking olefins in the
presence of napthenes, have reported a high ratio of saturates to
unsaturates among the products, showing that olefins are avid
hydrogen acceptors.

The transfer of hydrogen from some of the olefins [101,123]
absorbed on an acid surface will lead by sequential reactions to
dehydrogenated products which are not desorbable from the active
sites and in due course to the formation of coke on the catalyst.
There is no doubt that dehydrogenation is an important process
in coke formation in catalytic cracking.

Despite the importance of the hydrogen-transfer reaction, very
little is reported in the literature regarding its detailed mechanism

[124]. If we accept that the dehydrogenating carbenium ion exists on Brønsted site, it is hard to visualize how it will give up protons while preserving charge balance between the surface and the reactants. The same can be said for ions adsorbed on Lewis sites. To overcome this difficulty, one has to postulate one of two hypotheses:

1. The catalyst or coke is a conductor of $H^{\oplus}$ and/or $H^{\ominus}$, which flow to sites where hydrogen transfer is taking place in order to keep electrical neutrality on the surface.

2. Hydrogen transfer occurs between neutral species and hence by a free-radical process.

It is known that free-radical sites exist on acid catalysts and their participation in hydrogen-transfer reactions may be greater than suspected.

5.6.6 Coke Formation

In all reactions of hydrocarbons on acid catalysts there is a net production of a carbonaceous material called coke. Coke is an ill-defined material which is not desorbed from the catalyst upon purging with a suitable fluid, such as nitrogen or steam, over a defined length of time, at a given temperature. Such residues generally have a hydrogen-to-carbon ratio between 0.3 and 1.0 and have been reported to have spectroscopic features which suggest that bonding in these materials is similar to aromatic bonds. Extensive studies of coke have been conducted on its chemical origins, the nature of the coke produced, and its effect on the diffusion and activity properties of catalysts.

From the literature it appears that it is the condensation, alkylation, cyclization, and aromatization of materials present on the catalyst during catalytic cracking that eventually lead to the formation of coke by the transfer of hydrogen to gas-phase olefins [125,126]. Studies of the mechanism of carbon formation have utilized pure hydrocarbons which have shown that certain hydrocarbon species have a greater tendency than others for the formation of coke. Greensfelder and Voge [126] report that polynuclear aromatics, olefins, and polyolefins produce more carbon than naphthenes and paraffins. By feeding various types of hydrocarbons, such as paraffins, napthenes, olefins, and aromatics, they found that the structure of the resulting coke is similar in all cases. They also conclude that olefins form coke via an intermediate of an aromatic nature. In their studies a relationship between the formation of coke and the basicity of the various aromatic compounds was noted. Other authors [127] have reported that one of the most imported constituents found in coke involves condensed aromatic rings. To add to the confusion, these authors report that the nature of the feed has an influence on the nature of the coke formed.

It certainly appears that polyaromatics are important producers of coke. For instance, White [128] has related the tendency of a feedstock to form coke to its content of polyaromatics. At the same time, in the cracking of octane, it has been reported [123, 129] that the formation of coke is the result of hydrogen transfer from an absorbed species to an olefin and the polymerization and aromatization of alkyl carbenium ions on the surface, although how this can occur is not clear.

John et al. [116] report that on cracking a paraffinic gas oil completely free of polynuclear aromatics, the coke was a secondary product formed from olefins produced in the primary reactions of cracking. They also observed that no material of higher molecular weight than that of the materials contained in the feed were found in this reaction product. This suggests that coke is not the end product of a sequence of polymerization steps.

Hightower and Emmett [16] cracked n-hexadecane on silica alumina using radioactive tracers. They found that the relative amounts of radioactivity in the coke produced were as shown in Table 5.24.

From these results they conclude that olefins take part in the formation of coke but that paraffins and aromatics do not. It has been reported [130] that in the alkylation of benzene by ethylene, it was the ethylene that was responsible for the formation of coke in much greater proportion than benzene. Others [112,131] have found in studies of cumene cracking that detailed mass balances in the reaction reveal no loss of aromatic rings, and that all the coke formed comes from the side-chain carbons. This leads to the conclusion that it is the propyl carbenium ion or the product propylene which is responsible for all of the coke formation in this reaction.

Experiments with radioactive tracers in mixtures of 1-butene and decalin provide evidence on the relationship between the hydrogen-transfer ability of a catalyst and the formation of coke in silica aluminas [132]. In general, materials that have a higher

Table 5.24   Relative Coke-Forming Tendencies
of Various Hydrocarbons

Radioactive tracer	Relative amount of radioactivity in coke
n-Hexadecane	1.00 (base case)
Propane	0.07
Benzene	0.55
Toluene	1.71
Ethylene	1.75
Propylene	2.65
1-Pentene	8.58

Source: Ref. 16.

ability for hydrogen transfer produce less coke. This suggests
that hydrogen transfer is an important factor in the formation of
coke, although probably not the only factor [133]. Substantial
quantities of coke have been obtained on passing butadiene over
sodium-ammonium Y catalysts, and it has been postulated that the
formation of this coke is due to consecutive additions, of a Diels–
Alder type, on Lewis acids. The Brønsted acid sites are said to
contribute to the formation of coke by hydrogen-transfer reactions [82].

Langner has shown [134] that coke formation from propylene
on calcined sodium-ammonium Y and the deactivation of the catalyst
proceed by two different mechanisms, depending on the reaction
temperature. At reaction temperatures below 300°C the deactivation is caused mainly by the strong adsorption of compounds such
as cyclo-olefins and dienes in the pores of the zeolite catalyst.
Above 300°C aromatics are formed and can diffuse to the exterior
surface of the catalyst. Thus in considering the formation of
coke on catalysts, we must consider not only the nature of the
chemical reactions occurring but also the influence of diffusion
and hence of the mobility of the various species present in the
reacting mix. Indeed recently it has been reported that coke

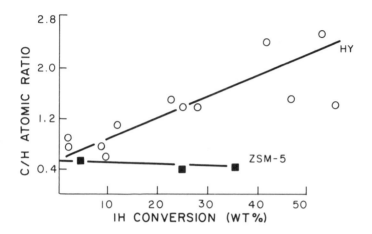

Figure 5.13   C/H ratio in coke as a function of conversion in the cracking of 1-hexene.   (From Ref. 86.)

formation is a spacially demanding reaction [135,136]; that is, in the case of zeolites it may be controlled by the nature of the pore structure.   Analysis of the coke formed in the reaction of ethylene on NaHY zeolites provides evidence that the number of fused or polynuclear aromatic rings is limited by the dimensions of the pore system of the catalyst [130].

The cracking of *n*-hexene-1 on HY and ZSM-5 has shown an interesting dependence of coke composition on catalyst type and the degree of conversion.   Figure 5.13 shows the observed C/H ratio in coke formed on these two catalysts.

The coke initially formed on both catalysts has a C/H ratio of about 0.5, which corresponds to that present in the feed.   As conversion increases, the coke on HY becomes more and more dehydrogenated, whereas on ZSM-5 the ratio does not change.   At the same time HY produces more saturated reaction products at high conversion than does the ZSM-5.

This observation confirms the difficulties associated with using coke on catalyst as a measure of catalyst decay and shows that under appropriate circumstances coke is both a product and a reactant in cracking.   How and why hydrogen is transferred from the coke to products on HY but not on ZSM-5 is far from clear. No doubt part of the explanation concerns the steric limitations present in the smaller pores of ZSM-5.   There remain the questions of how carbenium ions extract hydride ions from coke and how the residual charge is balanced. One is inclined to believe that the coke on the catalyst serves as a charge conductor, allowing the redistribution of charges generated, away from specific active sites on the inorganic matrix.

Other studies have found that most of the coke formed is not found inside the pores of the catalyst but is present on the exterior of the catalyst particles. In fact, some authors report that coke is found in the intersticies between particles of a cracking catalyst [137]. This agrees with investigations of the cracking of cumene on H mordenite, which show that deactivation of the catalyst is caused by an increase in the diffusional resistance between the particles of a catalyst pellet [138]. Furthermore, since some of the coke is known to have a graphitic structure, it cannot be accommodated inside the cavities of the catalyst and must be present on the outer surface of the catalyst particle.

From all of this it can be concluded that coke formation is not a well-understood process, not only because it is the result of a variety of reactions but also because the name "coke" encompasses an undetermined variety of nondesorbable species which appear on the catalyst during cracking. The general conclusion which can be drawn is that coke formation involves hydrogen transfer from surface species to gas-phase olefins. It seems that olefins are also the dominant species which adsorb on the surface, or polymerize by conjunct polymerization, and are the source of hydrogen used in saturating other olefins and the source of the carbon that remains as coke. Clearly, polynuclear aromatics and other heavy nondesorbable species will also contribute to the material known as coke. Light aromatics such as benzene and cumene do not seem to form coke from the aromatic ring itself. The majority of coke does not seem to occur on active sites themselves, but builds up on the exterior of the particle and in the intersticies between particles of a catalyst, as well as filling large pores in pellets of catalysts. The nature of the coke observed by the various investigators depends very much on its treatment before analysis, that is, on the purging conditions, temperature, nature of the purge gas, and the duration of the purge. Finally, the properties of the catalyst itself, such as the pore size and the geometry of the pores, seem to have an influence on the nature and amount of coke formed in a given reaction. Crystalline aluminosilicates such as ZSM-5 form very small amounts of coke, at least partly because the structure of the pore system is such that large molecules cannot be formed on the interior of the crystallite [135].

5.7   CRACKING OF MIXED ALKANES

Early workers who studied paraffin cracking tended to concentrate on long chains [7], whereas more recent work tends to concentrate on gasoline-range materials [33]. All such studies of pure components are essential to the proper understanding of the fundamentals of cracking. However, they are a long way from the type of complicated system that is presented by the cracking of commercial feeds. To bridge this gap it is instructive to study carefully composed mixtures of pure components in order to detect any synergistic interactions and to study the new phenomena that may arise in going from pure-component cracking to gas-oil cracking.

To this end, a study of mixed paraffins produced by Fischer—Tropsch synthesis seems ideally suited [139]. The feedstock consists of normal paraffins with a small impurity of monomethyl paraffins in their natural distribution as obtained from the synthesis. No olefins were present. In order to produce a usable feed, the synthesis product was cut at C_5 and C_{25} to produce the kind of material that could be used as a cracking feedstock in a plant producing synthetic gasoline [139].

It was observed that on cracking this feed, no C_1, C_2, or hydrogen appeared in the initial products of the reaction on HY. Primary products were limited to C_3 and above, with most of the material lying in the range C_3 to C_8. The product was highly isomerized, as shown in Figure 5.14, where a chromatogram of the light fraction of the feed is compared with the same fraction in the products. Very few olefins beyond C_7 were observed in the product and no significant isomerization occurred beyond this range.

Various observations lead to the following picture of the cracking of mixed long-chain paraffins on HY. Simple geometry indicates that long chains penetrate the zeolite catalyst pores linearly. As the chain penetrates it passes the first acid site in the pore and encounters its first opportunity to react. As each successive carbon atom of the chain passes the active site, the cumulative probability of reaction increases. From the observed product distribution one can conclude that by the seventh carbon or so, the cumulative probability is nearly 100%, resulting in a product distribution that is largely confined to C_8 and smaller molecules.

Consideration of the first cracking event shows that it must take place toward the long end of the molecule, as there are no olefins made that are longer than C_7. The short alkane produced by the initial protolysis is therefore the moiety which is deeper into the pore and at least partially blocked from backing out by

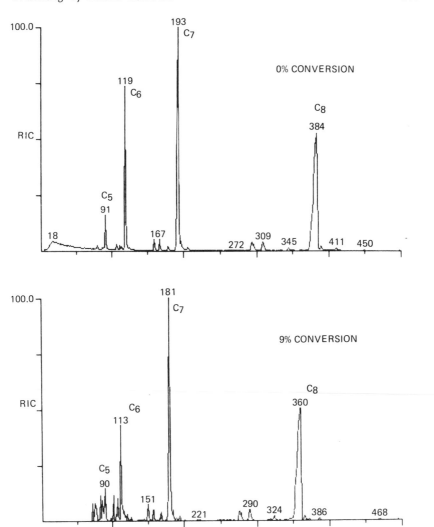

Figure 5.14 Chromatograms of the light fraction in FT products and the corresponding fraction after cracking of the mixed alkanes on HY. (From Ref. 139.)

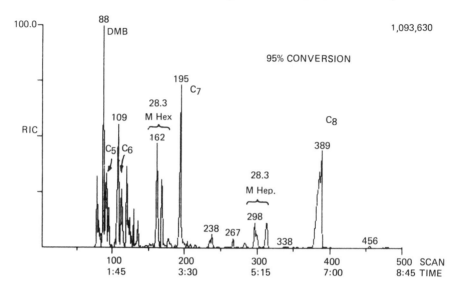

Figure 5.14   (Continued)

the residual carbenium ion on the site which has just catalyzed the scission.

To exit the pore structure, this short-chain molecule must execute a random walk through the interior of the crystallite seeking another exit. In this process it encounters many active sites and undergoes many reactions. Work on the reactions of short-chain molecules cited in Sections 5.2 and 5.3 shows that the majority of such reactions will be isomerizations and coking-saturation reactions. Indeed, the product of the cracking of mixed paraffins is highly isomerized and fairly well saturated.

The residual carbenium ion on the first site inside the pore can desorb as an olefin or continue to penetrate the pore by hydrogen shifts, in which case it soon cracks again, this time producing a short-chain olefin. No long-chain olefins are formed and very little isomerization, if any, takes place on the longer molecular chains, indicating that the adsorbed species continues to crack.

Such a mechanism explains all the results observed in the cracking of mixed paraffins on HY and leads to some unexpected conclusions.

1.   The HY zeolite crystallites are severely diffusion limited for the cracking of long-chain paraffins.

2. The olefins and short-chain paraffins that are produced initially are caught in a "cage effect" which forces them to undergo isomerization reactions and hence raises the octane level of the product.

3. Reduction of crystallite size should increase the cracking activity of HY but may also reduce isomerization and hence the octane number of the gaoline produced.

4. Coke deposition may be lower on smaller crystallites and olefin saturation is expected to be less pronounced.

5. There should be an optimum crystallite size in order to achieve the desired activity for cracking and isomerization.

It may well be that the zeolite crystallite size presently used is optimum for practical purposes, but it certainly would be a coincidence.

REFERENCES

1. P. A. Jacobs, Ed., *Carboniogenic Activity of Zeolites*, Amsterdam, Elsevier, 1977

2. A. Lopez Agudo, A. Asensio, and A. Corma, *J. Catal.*, *69*: 274 (1981); *Can. J. Chem. Eng.*, *60*: 50 (1982)

3. A. Corma, A. Lopez Agudo, I. Nebot, and F. Tomas, *J. Catal.*, *77*: 159 (1982)

4. F. LeNormand and F. Fajula, *Stud. Surf. Sci. Catal.* (Catal. Acids and Bases) (B. Imelik et al., Eds.), Amsterdam, Elsevier, p. 325, 1985

5. A. Corma, J. B. Monton, and A. V. Orchilles, *Ind. Eng. Chem.*, *Prod. Res. Dev.*, *23*: 404 (1984)

6. G. M. Pachenkov and A. S. Kazanskaya, *Zh. Fiz. Khim.*, *32*: 1779 (1958)

7. H. H. Voge, *Catalysis*, Vol. VI (P. H. Emmett, Ed.), p. 407, 1958

8. G. M. Good, H. H. Voge, and B. S. Greensfelder, *Ind. Eng. Chem.*, *39*: 1032 (1947)

9. J. H. Planelles, J. Sanchez—Marin, F. Tomas, and A. Corma, *J. Chem. Soc.*, *Perkin Trans. 2*, 333 (1985)

10. A. Corma, J. H. Planelles, J. Sanchez—Marin, and F. Tomas, *J. Catal.*, *92*: 284 (1985)

11. B. S. Greensfelder and H. H. Voge, *Ind. Eng. Chem.*, *37*: 514 (1945)

12. B. S. Greensfelder, H. H. Voge, and G. M. Good, *Ind. Eng. Chem.*, *41*: 2573 (1949)

13. W. A. Van Hook and P. H. Emmett, *J. Am. Chem. Soc.*, *84*: 4410 (1962)

14. A. Van Hook and P. H. Emmett, *J. Am. Chem. Soc.*, *85*: 697 (1963)

15. J. L. Bordley, Jr. and P. H. Emmett, *J. Catal.*, *42*: 367 (1976)

16. J. W. Hightower and P. H. Emmett, *J. Am. Chem. Soc.*, *87*: 939 (1965)

17. P. E. Pickert, J. A. Rabo, E. Dempsey, and V. Schomaker, *Proc. 3rd Int. Congr. Catal.*, *Amsterdam*, *1*: 714 (1964)

18. S. E. Tung and E. McIninch, *J. Catal.*, *10*: 166 (1968)

19. J. N. Miale, N. Y. Chen, and P. B. Weisz, *J. Catal.*, *6*: 278 (1966)

20. H. Schulz and A. Geertsema, *Proc. 5th Int. Conf. Zeolites* (V. L. Rees, Ed.), London, Heyden, p. 874, 1980

21. D. K. Thakur and S. W. Weller, *Adv. Chem. Ser.* (Mol. Sieves, Int. Conf., 3rd), *101*: 596 (1973)

22. G. Lopez, G. Perot, C. Gueguen, and M. Guisnet, *Acta Phys. Chem.*, *24*: 207 (1978)

23. P. D. Hopkins, *J. Catal.*, *12*: 325 (1968)

24. L. P. Aldridge, J. R. McLaughlin, and C. G. Pope, *J. Catal.*, *30*: 409 (1973)

25. G. Pop, G. Musca, E. Pop, P. Tomi, and G. Ivanus, *Rev. Chim. Bucharest*, *32*: 317 (1981)

26. D. A. Best and B. W. Wojciechowski, *J. Catal.*, *47*: 11 (1977)

27. G. A. Fuentes and B. C. Gates, *J. Catal.*, *76*: 440 (1982)

28. A. P. Bolton and M. A. Lanewala, *J. Catal.*, *18*: 1 (1970)

29. P. B. Weisz and J. N. Miale, *J. Catal.*, *4*: 527 (1965)

30. R. Maatman, C. Friesema, R. Mellema, and J. Maatman, *J. Catal.*, *47*: 62 (1977)

31. A. N. Ko and B. W. Wojciechowski, *Prog. React. Kinet.*, *12*: 201 (1984)

32. A. Corma and B. W. Wojciechowski, *Catal. Rev. Sci. Eng.*, *24*: 1 (1982); *Prepr. Div. Pet. Chem.*, *Am. Chem. Soc.*, *28*: 861 (1983)

33. D. Cornet and A. Chambellan, *Stud. Surf. Sci. Catal.* (Catal. Acids and Bases) (B. Imelik et al., Eds.), Amsterdam, Elsevier, p. 273, 1985

34. R. Beaumont, D. Barthomeuf, and Y. Trambouze, *Adv. Chem. Ser.* (Mol. Sieve Zeolites), *102*: 327 (1971)

35. P. B. Venuto and P. S. Landis, *Adv. Catal.*, *18*: 259 (1968)

36. T. Yashima, A. Yoshimura, and S. Namba, *Proc. 5th Int. Conf. Zeolites, Naples* (L. V. Rees, Ed.), London, Heyden, p. 705, 1980

37. N. Y. Chen and P. B. Weisz, *Chem. Eng. Prog. Symp. Ser.*, *63*(73): 86 (1967)

38. S. M. Csicsery, ACS Monograph 171, *Zeolite Chem. Catal.* (J. A. Rabo, Ed.), p. 680, 1976

39. C. Naccache and Y. Ben Taarit, *Proc. 5th Int. Conf. Zeolites. Naples* (L. V. Rees, Ed.), London, Heyden, p. 592, 1980

40. E. G. Derouane, *Stud. Surf. Sci. Catal.* (Catal. Zeolites) (B. Imelik et al., Eds.), Amsterdam, Elsevier, p. 5, 1980

41. K.-P. Wendlandt, W. Weigel, F. Hofmann, and H. Bremer, *Z. Anorg. Allg. Chem.*, *445*: 51 (1978)

42. K.-P. Wendlandt, W. Franke, H. Bremer, K. Becker, and K.-H. Steinberg, *Z. Anorg. Allg. Chem.*, *445*: 59 (1978)

43. R. M. Barrer and D. A. Harding, *Separ. Sci.*, *9*: 195 (1974)

44. T. E. Whyte, Jr., E. L. Wu, G. T. Kerr, and P. B. Venuto, *J. Catal.*, *20*: 88 (1971)

45. J. A. Gard and J. M. Tait, *Adv. Chem. Ser.* (Mol. Sieve Zeolites—I), *101*: 230 (1971)

46. N. Y. Chen, *Proc. 5th Int. Congr. Catal.*, Amsterdam, *2*: 1343 (1973)

47. C. Mirodatos and D. Barthomeuf, *J. Catal.*, *57*: 136 (1979)

48. N. Y. Chen, S. J. Lucki, and E. B. Mower, *J. Catal.*, *13*: 329 (1969)

49. R. L. Gorring, *J. Catal.*, *31*: 13 (1973)

50. N. Y. Chen and W. E. Garwood, *Adv. Chem. Ser.* (Mol. Sieves, Int. Conf., 3rd), *121*: 575 (1973)

51. M. S. Spencer and T. V. Whittan, *Spec. Publ.—Chem. Soc. Properties and Applications of Zeolites* (R. P. Townsend, Ed.), p. 342, 1980

52. J. Dewing, F. Pierce, and A. Stewart, *Stud. Surf. Sci. Catal.* (Catal. Zeolites) (B. Imelik et al., Eds.), *5*: 39, 1980

53. W. E. Garwood and N. Y. Chen, *Prepr.*, *Div. Pet. Chem.*, *Am. Chem. Soc.*, *25*: 84 (1980)

54. N. Y. Chen and W. E. Garwood, *J. Catal.*, *52*: 453 (1978)

55. S. J. Miller, *Belg. Patent 879,156*, 1980

56. S. J. Miller and T. R. Hughes, *U.S. Patent 4,190,519*, 1980

57. D. J. O'Rear and J. F. Mayer, *Belg. Patent 877,772*, 1979

58. D. J. O'Rear and J. F. Mayer, *U. S. Patent 4,171,257*, 1979

59. G. A. Olah, *J. Am. Chem. Soc.*, *94*: 808 (1972)

60. G. A. Olah, J. R. DeMember, and J. Shen, *J. Am. Chem. Soc.*, *95*: 4952 (1973)

61. G. A. Olah, J. Shen, and R. H. Schlosberg, *J. Am. Chem. Soc.*, *95*: 4957 (1973)

62. G. A. Olah, Y. Halpern, J. Shen, and Y. K. Mo, *J. Am. Chem. Soc.*, *95*: 4960 (1973)

63. G. A. Olah, Y. K. Mo, and J. A. Olah, *J. Am. Chem. Soc.*, *95*: 4939 (1973)

64. H. Hattori, O. Takahashi, M. Takagi, and K. Tanabe, *J. Catal.*, *68*: 132 (1981)

65. O. Takahasi and H. Hattori, *J. Catal.*, *68*: 144 (1981)

66. D. M. Nace, *Ind. Eng. Chem.*, *Prod. Res. Dev.*, *8*: 31 (1969)

67. A. Borodzinski, A. Corma, and B. W. Wojciechowski, *Can. J. Chem. Eng.*, *58*: 219 (1980)

68. B. C. Gates, J. R. Katzer, and G. C. A. Schuit, *Chemistry of Catalytic Processes*, New York, McGraw-Hill, 1979

69. W. O. Haag and R. M. Dessau, *Proc. 8th Int. Congr. Catal. Berlin*, 2: 305 (1984)

70. B. S. Greensfelder, H. H. Voge, and G. M. Good, *Ind. Eng. Chem.*, 49: 747 (1957)

71. P. B. Janardhan and S. Rajeswari, *Ind. Eng. Chem.*, *Prod. Res. Dev.*, 16: 52 (1977)

72. A. Brenner and P. H. Emmett, *J. Catal.*, 75: 410 (1982)

73. P. B. Weisz, *Ann. Rev. Phys. Chem.*, 21: 175 (1970)

74. J. Scherzer and R. E. Ritter, *Ind. Eng. Chem.*, *Prod. Res. Dev.*, 17: 219 (1978)

75. A. P. Bolton and R. L. Bujalski, *J. Catal.*, 23: 331 (1971)

76. P. Aldridge, J. R. McLaughlin, and C. G. Pope, *J. Catal.*, 30: 409 (1973)

77. D. M. Anufriev, P. N. Kuznetlov, and K. G. Ione, *React. Kinet. Catal. Lett.*, 9: 297 (1978)

78. P. B. Weisz, *Chem. Technol.*, 3: 498 (1973)

79. H. Pines and R. C. Wackher, *J. Am. Chem. Soc.*, 68: 595 (1946)

80. V. L. Magnotta and B. C. Gates, *J. Catal.*, 46: 266 (1977)

81. V. Haensel, *Adv. Catal.*, 3: 179 (1951)

82. B. E. Langner and S. Meyer, *Stud. Surf. Sci. Catal.* (Catal. Deact.) (B. Delmon and G. F. Froment, Eds.), 6: 91, 1980

83. B. V. Liengme and W. K. Hall, *Trans. Faraday Soc.*, 62: 3229 (1966)

84. N. W. Cant and W. K. Hall, *J. Catal.*, 25: 161 (1972)

85. J. Abbot and B. W. Wojciechowski, *Can. J. Chem. Eng.*, 63: 462 (1985)

86. J. Abbot and B. W. Wojciechowski, *Can. J. Chem. Eng.*, 63: 818 (1985)

87. A.-N. Ko and B. W. Wojciechowski, *Int. J. Chem. Kinet.*, 15: 1249 (1983)

88. J. Abbot, A. Corma, and B. W. Wojciechowski, *J. Catal.*, 92: 398 (1985); J. Abbot and B. W. Wojciechowski, *J. Catal.*, 90: 270 (1984)

89. D. W. Basset and H. W. Habgood, *J. Phys. Chem.*, *64*: 769 (1960)

90. W. K. Hall, F. E. Lutinski, and H. R. Garberich, *J. Catal.*, *3*: 512 (1964)

91. T. Ishii and G. L. Osberg, *AIChE J.*, *11*: 279 (1965)

92. W. R. Stevens and R. G. Squires, *Chem. React. Eng.*, *Proc. 5th European Symp.*, *B2*: 35 (1972)

93. J. Weitkamp and S. Ernst, *Stud. Surf. Sci. Catal.* (Catal. Acids and Bases) (B. Imelik et al., Eds.), Amsterdam, Elsevier, p. 419, 1985

94. B. D. Flockhart, L. McLoughlin, and R. C. Pink, *J. Chem. Soc.*, *D*, 818 (1970)

95. P. T. Wierzchowski, S. Malinowski, and S. Krzyzanowski, *Chim. Ind. (Milan)*, *59*: 612 (1977)

96. P. Fejes, I. Hannus, I. Kiricsi, and K. Varga, *Acta Phys. Chem.*, *24*: 119 (1978)

97. N. T. Tam, R. P. Cooney, and G. Curthoys, *J. Catal.*, *44*: 81 (1976)

98. I. Kiricsi, I. Hannus, K. Varca, and P. Fejes, *J. Catal.*, *63*: 501 (1980)

99. A. Corma and A. Lopez Agudo, *React. Kinet. Catal. Lett.*, *16*: 253 (1981)

100. V. V. Kharlamov, T. S. Starostina, and Kh. M. Minachev, *Izv. Akad. Nauk SSSR*, *10*: 2291 (1982)

101. M. E. Guzman, C. Parra, and M. R. Goldwasser, *Acta Cient. Venezuela*, *32*: 495 (1981)

102. V. Illes and H. Schindlbauer, *Acta Chim. Acad. Sci. Hung.*, *82*: 449 (1974)

103. Kh. Dimitrov, *God. Sofii. Univ. Fiz.-Mat. Fak.*, *51*: 155 (1959)

104. P. B. Venuto, L. A. Hamilton, P. S. Landis, and J. J. Wise, *J. Catal.*, *5*: 81 (1966)

105. H. G. Karge, J. Ladebeck, and Z. Sarbak, *Stud. Surf. Sci. Catal.*, *7*: 1408 (1981)

106. H. G. Karge, J. Ladebeck, Z. Sarbak, and K. Hatada, *Zeolites*, *2*: 94 (1982)

107. A. Corma and B. W. Wojciechowski, *Catal. Rev. Sci. Eng.*, 24: 1 (1982)

108. C. D. Prater and R. M. Lago, *Adv. Catal.*, 8: 293 (1956)

109. Y. Murakami, T. Hattori, and T. Hatton, *J. Catal.*, 10: 123 (1968)

110. W. F. Pansing and J. B. Malloy, *Ind. Eng. Chem.*, Process Des. Dev., 4: 181 (1965)

111. J. T. Richardson, *J. Catal.*, 9: 182 (1967)

112. D. A. Best and B. W. Wojciechowski, *J. Catal.*, 47: 11 (1977)

113. A. Corma and B. W. Wojciechowski, *J. Catal.*, 60: 77 (1979)

114. A. Corma and B. W. Wojciechowski, *Can. J. Chem. Eng.*, 58: 620 (1980)

115. D. R. Campbell and B. W. Wojciechowski, *J. Catal.*, 20: 217 (1971)

116. T. M. John, R. A. Pachovsky, and B. W. Wojciechowski, *Adv. Chem. Ser.*, 133: 422 (1974)

117. P. D. Bartlett, F. E. Condon, and A. Schereider, *J. Am. Chem. Soc.*, 66: 1531 (1944)

118. F. E. Condon, *Catalysis*, Vol. VI (P. H. Emmett, Ed.), New York, Reinhold, p. 43, 1958

119. H. Pines, *Chemistry of Catalytic Hydrocarbon Conversions*, New York, Academic Press, 1981

120. J. C. Vedrine, P. Dejaifve, C. Naccache, and E. G. Derouane, *Proc. 7th Int. Congr. Catal.*, Tokyo, 7: 724 (1980)

121. C. D. Nenitzescu, *Carbonium Ions* (G. A. Olah and P. R. Schleyer, Eds.), Vol. II, p. 463, New York, Wiley, 1970

122. H. H. Voge, G. M. Good, and B. S. Greensfelder, *Ind. Eng. Chem.*, 38: 1033 (1946)

123. C. L. Thomas, *J. Am. Chem. Soc.*, 66: 1586 (1944)

124. C. F. Parra, M. R. Goldwasser, and F. Figueras, *II Colloque Franco—Venezuelien de Catal.*, IFP ed. D-1, 1985

125. P. B. Venuto, *Chem. Tech.*, 1: 215 (1971)

126. B. S. Greensfelder and H. H. Voge, *Ind. Eng. Chem.*, *37*: 983 (1945); *37*: 1038 (1945); B. S. Greenfelder, H. H. Voge, and G. M. Good, *Ind. Eng. Chem.*, *37*: 1168 (1945)

127. P. E. Eberly, Jr., C. N. Kimberlin, W. H. Miller, and H. V. Drushel, *Ind. Eng. Chem.*, *Process Des. Dev.*, *5*: 193 (1966)

128. P. J. White, *Hydrocarbon Process*, *47*(5): 103 (1968)

129. A. A. Petrov and A. V. Frost, *Dokl. Akad. Nauk SSSR*, *65*: 851 (1949)

130. P. B. Venuto and L. A. Hamilton, *Ind. Eng. Chem.*, *Prod. Res. Dev.*, *6*: 190 (1967)

131. C. C. Lin, S. W. Park, and W. Hatcher, *Ind. Eng. Chem.*, *Process Des. Dev.*, *22*: 609 (1983)

132. R. W. Blue and C. J. Engle, *Ind. Eng. Chem.*, *43*: 494 (1951)

133. D. E. Walsh and L. D. Rollman, *J. Catal.*, *49*: 369 (1977)

134. B. E. Langner, *Ind. Eng. Chem.*, *Process Des. Dev.*, *20*: 326 (1981)

135. L. D. Rollman, *J. Catal.*, *47*: 113 (1977)

136. L. D. Rollman and D. E. Walsh, *J. Catal.*, *56*: 139 (1979)

137. R. G. Haldeman and M. C. Botty, *J. Phys. Chem.*, *63*: 489 (1958)

138. J. B. Butt, *J. Catal.*, *41*: 190 (1976)

139. J. Kobalakis and B. W. Wojciechowski, *Can. J. Chem. Eng.*, *63*: 269 (1985); J. Abbot and B. W. Wojiechowski, *Ind. Eng. Chem.*, *Prod. Res. Dev.*, *24*: 501 (1985)

140. M. Marczewski and B. W. Wojciechowski, *Can. J. Chem. Eng.*, *60*: 617 (1982)

6

Catalytic Cracking of Gas Oils

6.1 REVIEW OF THE PHENOMENA INVOLVED IN CRACKING

Before proceeding with the examination of the catalytic cracking of gas oils, one should stop and examine what has been ascertained from work on the simpler systems described in Chapter 5.

There is no doubt that carbocations are the principal, if not the sole agents responsible for catalytic cracking and associated events. It appears that both carbenium and carbonium ions are involved. Both are formed on Brønsted sites, whereas Lewis sites produce only carbenium ions.

The acid strength distribution as well as the Brønsted/Lewis ratio will have an effect on both the activity (as measured by conversion at standard conditions) and the selectivity exhibited by various catalyst formulations without influencing the nature of the individual reactions that occur in cracking. The observed selectivity changes on various catalysts are due to changes in the rates of individual reactions caused by differences in appropriate site concentrations, not in the nature of the sites themselves. It seems likely that each reaction proceeds on sites with their own narrow range of acidities and their own nature (i.e., Brønsted or Lewis.

Since the rates of reactions are dependent on site densities, while the activation energies for the various reactions vary over a range of some 25 to 30 kcal/mol, selectivity can be altered over a very wide range by a combination of reaction temperature and catalyst formulation.

All cracking catalysts decay, and the most active decay most rapidly. The decay introduces phenomena which are not intuitively obvious to those accustomed to classical kinetics and catalysts. These phenomena have been quantified and can be well described by the time-on-stream theory. Application of the theory has made clear many of the puzzling features of catalytic cracking, not the least of which is the sharp and confusing difference in the observed selectivity obtained in experimental confined-bed test reactors and commercial steady-state reactors. Use of the theory has been of great help in defining unambiguously the networks of initial reactions in pure compound cracking, as well as in allowing rational extrapolation of catalyst behavior to "fresh catalyst" conditions. This allows the nature of the active sites to be studied by kinetics at accessible reaction conditions.

Studies of mixed-feed reactions have led to ambiguous results. On the one hand, the addition of an olefin to an alkane is reported to increase the cracking of the alkane; on the other, a mixture of alkanes shows no signs of synergystic effects in cracking. Whatever synergystic effects there may be, they have not drawn much attention to date. Certainly, the various molecular species present in a mixed feed, as well as their products, interact in secondary reactions and occupy the same active sites. Whether that can be called synergysm is a question of definition.

We can be sure that the phenomena that will present themselves in the cracking of the highly complex materials that constitute a gas oil will obey all of the fundamental behaviors mentioned above, with an added overlay of complexity due to the complexity of the feed.

6.2   PRODUCT DISTRIBUTION IN GAS—OIL CRACKING

According to the theoretical considerations presented in the earlier chapters, one could expect that the primary products in the cracking of paraffinic gas oil will contain no hydrogen, nor any C_1 or C_2 molecules, and will include propylene, butenes, olefins, and paraffins, all of shorter chain length than the original feedstock. One would not expect products of molecular weight larger than the original gas oil. Nonetheless, early workers reported 2 to 5 wt % of C_1 and C_2 hydrocarbons in the cracking products [1]. This result is not in agreement with the carbenium ion theory, and some have ascribed their appearance to "less favored" types of cracking reactions. Much, if not all, of this type of product is due to competitive thermal cracking which occurs in the space between catalyst particles at the temperatures involved in catalytic cracking. The rest of the C_1 and C_2 product can be ascribed to secondary reactions of the primary products of the catalytic reaction. Detailed studies of the catalytic cracking of selected gas oils have shown that in fact the carbenium ion theory is well obeyed by the distribution of the primary products of cracking on ultrastable HY catalysts [2].

The selectivity curves shown in Figures 6.1 and 6.2 show that propane, propylene, *n*-butane, butenes, and gasoline are the primary products of the catalytic cracking of a paraffinic gas oil, while methane, ethane, ethylene, and coke are clearly secondary products. In these figures the dashed lines show the yields of the various products obtained in static bed runs at a fixed catalyst-to-oil ratio at various times on stream. The enveloping solid lines correspond to the OPE, that is, the selectivity for the given product which would be obtained at zero time on stream or, in other words, on completely fresh catalyst. All primary products can in turn interact with the acid centers of the catalyst, giving rise to the formation of carbenium ions which lead to the appearance of secondary products. For example, an olefin formed in a primary reaction can be adsorbed on the surface

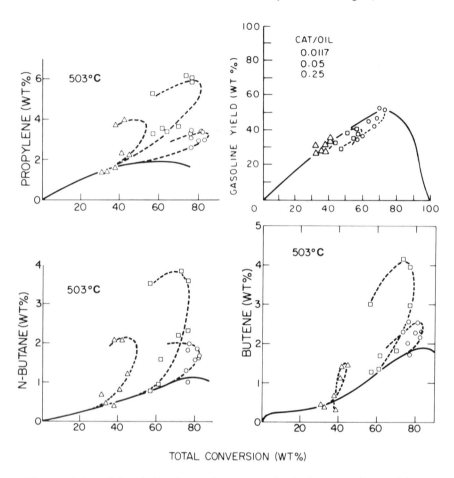

Figure 6.1   Selectivity for primary products in gas-oil cracking.
(From Ref. 2.)

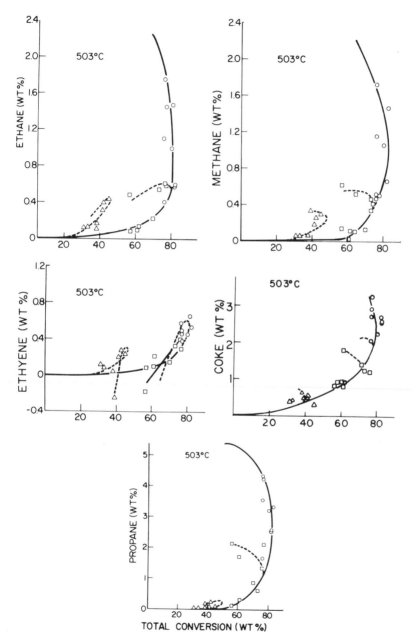

Figure 6.2   Selectivity for secondary products in gas-oil cracking.   (From Ref. 2.)

of the catalyst and give up a hydrogen to another olefin in the
gas phase, leading to the formation of a paraffin and a coke precursor. Such reactions, which enrich certain molecules in hydrogen content at the expense of others, are encouraged by the
hydrogen-transfer properties of the catalyst.

In a desirable cracking catalyst, the cracking of low-molecular-
weight materials will be slow, while the cracking of large molecules to materials falling in the range of gasoline will be relatively
fast. Since the cracking of large molecules is easier than the
cracking of small molecules, a desirable catalyst has an acid
strength distribution which is just right for the cracking of large
molecules but too weak for the cracking of small molecules. Furthermore, a successful cracking catalyst will not transfer hydrogen
too readily [3] since such reactions lead to the formation of coke
and consequent deactivation of the catalyst. At the same time,
the octane number of the gasoline produced is greatly enhanced
by the presence of olefins and isomerized aliphatics. A desirable
catalyst should therefore preserve the olefins while isomerizing
straight chains to the branched isomeric forms. Hydrogen transfer will saturate the shorter-chain olefins formed, giving the corresponding paraffins, which are difficult to crack. This in turn
will stabilize the gasoline fraction, decreasing the rate of overcracking.

That methane, ethane, and ethylene appear as products only at
high conversions attests either to the fact that they are tertiary
or later products of catalytic cracking, or to the fact that they
are formed by thermal reactions operating on the primary products
of catalytic cracking. Unfortunately, there is very little information in the literature on reactions that produce the light gases as
genuine primary products.

The formation of isobutane as a primary product is not easy to
explain by one-step conventional cracking mechanisms. Various
authors [5] have found isoparaffins as primary products in the
cracking of n-hexadecane and of methylcyclohexane. Mechanisms
that explain the presence of methylalkanes in primary products
assume that chain isomerization takes place very rapidly before
the carbenium ion can be desorbed [15]. A postulate which needs
more examination is that the short-chain products are isomerized
due to repeated secondary reactions inside the pore system of the
zeolite, as per the "cage effect" described previously.

Coke is an unavoidable product of catalytic cracking. The ac-
cumulation of coke on the catalyst leads to catalyst deactivation
while providing hydrogen for the saturation of olefins. Both
those effects are largely undesirable in catalytic cracking, and it
is therefore of some considerable interest to examine the rates

and sources of coke formation. The one desirable facet of coke formation is that in practice the combustion of coke in the regenerator provides the heat of reaction for catalytic cracking in commercial processes.

From Figure 6.2 it can be seen that here as well as in the case of the gas oil used by John and Wojciechowski [6], coke is a secondary product of the cracking reaction. Eberly et al. [7] arrived at the same conclusion by studying the concentration of coke as a function of catalyst position in a fixed-bed reactor in the cracking hexadecane. These authors observed the carbon profile to be displaced to the lower part of a fixed catalyst bed as the space velocity of the feed was increased. It was concluded that coke was formed from olefins that are produced in the primary cracking reactions. At the same time, if the feed contains polyaromatic components, the formation of coke is found to be a primary reaction, although it is not entirely clear whether the coke thus formed is a product of a catalytic reaction or merely represents the adsorbed polynuclear aromatics of the feed in a dehydrogenated form. Walsh and Rollman, for instance, have shown [8] that if one feeds a mixture of paraffins and complex aromatics over a cracking catalyst, the coke is produced mainly from the aromatics.

Since the detailed chemistry of catalytic cracking of gas oil is intractable at present, one is forced to examine generalities rather than details. One of the methods used to do this is to consider the behavior of groups of compounds as a unit. In this way light can be shed on both the product distribution and the kinetics of gas-oil cracking.

6.3   KINETICS OF GAS—OIL CRACKING

6.3.1   Reaction Networks

In view of the fact that the gas-oil feed contains many types of molecules, attempts have been made to group various similar components into a few "cuts" or "lumps." Using this concept of lump groupings, Weekman and others [9,10] present the reaction scheme shown in Figure 6.3. Using these lumps, they have formulated a kinetic model to describe gas-oil cracking. Their model also takes into account deactivation and the effects of nitrogenous and polyaromatic compounds.

In similar systems Wojciechowski and colleagues [6,30,33] have reported the reaction network shown in Figure 6.4 based on their

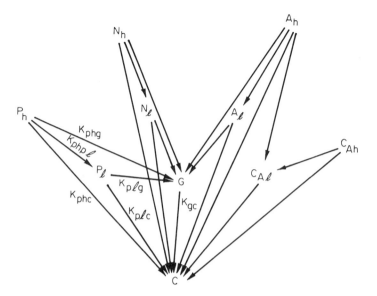

Figure 6.3 Lumped kinetic scheme for gas-oil cracking. (From Ref. 10.)

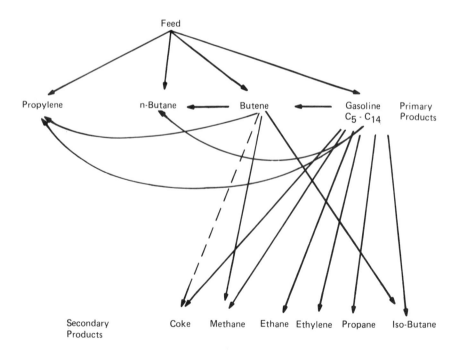

Figure 6.4 Reaction scheme for catalytic cracking of gas oils. (From Ref. 6.)

studies of the cracking of various gas oils containing no poly-nuclear aromatics.

6.3.2   Influence of Feedstock Composition

It has been suggested that in a multicomponent feed the mixture does not behave in a way that can be understood by an investiga-tion of each of the hydrocarbons in isolation. The behavior of such a mixture of hydrocarbons may involve a variety of inter-actions [11,12] which are not possible when the hydrocarbons are studied one by one. If this is so, it will lead to a great deal of difficulty in the interpretation of the results observed when crack-ing mixed feeds and offers no hope that a catalog of the behaviors of individual components or lumps of components will provide us with an adequate picture of the behavior of a mixture of these components. Nevertheless, parameters have been proposed that can be used to characterize the feedstock and to predict yields in mixed-feed reactions. Watson and Nelson [13] have proposed an empirical "index" for the nature of the feedstock which they ex-press by the factor F:

$$F = \left[\frac{\text{average molal boiling point (°R)}}{\text{specific gravity at 60°F}} \right]^{1/3} \tag{6.1}$$

F is found to range from 12.5 for purely paraffinic feeds to 10 or less for feeds including recycle from the reactor. The F index has been correlated with the cracking rate of certain feedstocks and with the octane number of the gasoline obtained [14]. The correlation is strictly empirical and similar to related correlations proposed in the past [15].

Subsequently, using the NDM method, which permits identifica-tion of the structures of paraffins, napthenics, and aromatics in a mixture of hydrocarbons, Reif et al. [16] developed a polynomial expression correlating the yield of various products with the weight percentages of paraffins, napthenes, and aromatics in the feed. This correlation takes the form

$$\text{yield} = f(\text{MABP}, C_p, C_N, C_A) \tag{6.2}$$

in which MABP is the medium average boiling point (°R), and C_p, C_N, and C_A correspond, respectively, to the atom percentages of carbon present in paraffinic, napthenic, and aromatic compounds in the feed.

This correlation shows that, for example, if one keeps the percentage of napthenes constant while increasing the paraffins in the feed, one increases the coke formation while decreasing the formation of light gases. It is also found that in a feed composed of a mixture of paraffins and napthenes, if one varies the relative proportion of the two components, one varies the production of gases, keeping carbon formation constant. This strictly empirical correlation has no fundamental significance but can be of great practical utility in guiding feed blending for commercial cracking operations. Furthermore, it offers an indication that the various components crack independently.

Further advances in analytical techniques [17] have permitted the classification of feeds on the basis of nine major types of constituents. This classification allows one to differentiate between mono- and polycyclic aromatics, paraffins, olefins, and other constituents in order to construct a correlation that can be used to predict the distribution of products which are highly dependent on the nature of the original feed molecules. For example, monocyclic compounds produce high yields of gasoline and low yields of coke, whereas polyaromatics give almost directly opposite results [18].

Nace and others [19,20] confronted the problem of feed composition with rigor and exactitude. They studied a large number of feeds and examined the influence of feed composition on the kinetic rate constant and on the deactivation of the catalysts. They report that increasing the ratio of paraffins to napthenes leads to a decrease in the rate of cracking of the gas oil, to a decrease in the rate of gasoline formation and the rate of gasoline recracking, and to an increase in the rate of catalyst deactivation. They also report that whereas the rate of reaction can vary by a factor of 4 in the range of compositions studied, gasoline selectivity did not seem to vary substantially. Unfortunately, they do not report a relationship between the kinetic parameters obtained in this way and any of the classes of hydrocarbons present in the feedstock [21].

On reflection it becomes obvious that it is not enough to take into account the composition [22] of the initial feedstock used in catalytic cracking, since the composition of the material presented to the catalyst varies with conversion during the cracking reaction. For example, it is clear that if one begins by cracking a purely paraffinic feed in a fixed-bed reactor, soon after encountering the first increment of catalyst the feed becomes mixed with olefins that arise as primary products of cracking in the first portion of the reactor. Subsequent catalyst increments are exposed to a mixed feed consisting of paraffins and olefins, thus complicating the issue of the composition of the feedstock and of how the observations should be treated. The composition clearly varies along the bed length and depends on the level of conversion. Furthermore, the olefins produced in the primary reaction are readily isomerized and may be converted to napthenes and aromatics, further complicating the picture in subsequent layers of a fixed-bed reactor.

In the case of a commercial riser or moving-bed reactor, as the gas oil traverses the cracker, the components that are most readily converted are converted in the initial portion of the reactor. Subsequent portions of the reactor are presented not only with the products of the initial reaction but also with the more and more refractory residue of the original feed. This explains the well-known observation that the cycle stock of catalytic crackers is much more difficult to crack than the original feed. It is reported that very high catalyst-to-oil ratios have to be used to achieve appreciable conversion of cycle stock [23] that contains the refractory residue of the original feed. Moreover, because of the recycling of the unconverted feed, the feed composition in commercial gas-oil cracking varies with the level of conversion.

The effect of this was quantified by Kemp and Wojciechowski [24], who derived the following equation for the rate of gas-oil cracking, taking into account the mixed-feed nature of the reactant:

$$-\frac{dC_A}{d\tau} = k_0 C_s C_A \sum_{i=1}^{\infty} k_{i0} X_i \qquad (6.3)$$

where the k_{i0} are related to the rate constants for the cracking of individual compounds, C_S and C_A are site and overall feed concentrations, X_i is the mole fraction of the ith component, and k_0 is the "overall initial rate constant."

In Equation (6.3) the term $\Sigma\, k_{i0}X_i$ accounts for the variation in the observed rate constant with composition changes. The authors show that this term can be very well approximated under a wide range of conditions by the expression $(C_A/C_{AO})^W$. The new parameter, W, which they call the "refractoriness" parameter, can be used to quantify variations between feedstocks of different refractoriness (i.e., ones that show different levels of resistance to cracking with depth of conversion). The authors report two other cases where the modeling function shown above is inadequate to describe refractoriness and suggest appropriate model expressions for those cases. Fortunately, in all the catalytic cracking work reported to date, the simple model accounts very well for the refractoriness observed and has been applied successfully to a large number of feedstock/catalyst combinations [2,13,25].

6.3.3   Kinetic Models

One of the early attempts to describe the catalytic conversion of gas oil by a kinetic model was that due to Voorhies [26]. According to this author, for a given catalyst and a given feedstock at one temperature, there is a good correlation between the conversion obtained and the yield of coke deposited on the catalyst. In this way an expression giving average conversion as a function of coke on catalyst for a fixed-bed reactor was developed in the following form:

$$\overline{X} = \left(\frac{A'_c}{A'_f F}\right)^{1/N_f} t^{(N_c - 1)/N_f} \tag{6.4}$$

where $\overline{X}$ is the average conversion over some period of time; A'_c, A'_f, N_c, and N_f are parameters; t is the time on stream; and F is the weight hourly space velocity. Voorhies observed that the weight of carbon deposited was uniform with time of reaction for a given catalyst, feed composition, and series of reaction conditions. By studying a large number of experimental results, he concluded that the amount of carbon deposited on the catalyst was proportional to the logarithm of the time on stream:

$$C_c = A'_c t^{N_c} \tag{6.5}$$

From this relationship one could calculate the parameters A'_c and N_c. To calculate the parameters A_f and N_f, an empirical relationship was established between the percentage of carbon on feed and the average conversion:

$$C_f = A'_f \bar{X}^{N_f}$$

(6.6)

The Voorhies relationship has been shown to fit some experimental results very well and continues to be used as a useful correlation for limited ranges of conditions.

Blanding [27] was the first to develop a mathematical model based on a kinetic rate expression. He assumed that the cracking of hydrocarbons follows a first-order rate law. On the basis of experimental data he changed his rate expression to a second-order form to account for volume expansion on cracking. This second-order model yields instantaneous fractional conversions in a fixed-bed reactor in the form

$$\frac{k_i \pi}{F} = \frac{X}{1 - X}$$

(6.7)

where π is the total pressure, F the volumetric feed rate, and k_i the instantaneous rate constant for cracking. Realizing that catalyst activity varies with time on stream, Blanding defined an average rate constant for cracking as follows:

$$k = \frac{1}{t} \int_0^t k_i \, dt$$

(6.8)

Using the modified rate constant, an integral model of catalytic cracking of gas oil was formulated:

$$\frac{k \pi (\rho_c / \rho_o)}{S} = \frac{\bar{X}}{1 - \bar{X}}$$

(6.9)

where S is the volumetric flow rate; ρ_c and ρ_0 are the densities of catalyst and gas oil, respectively, and $\overline{X}$ is the time-averaged fraction converted.

Blanding's equation fits his data well enough within a limited range of experimental conditions, but since it lacks a relationship between the instantaneous rate constants and the time on stream, it is necessary to make a large number of runs in order to define instantaneous rate constants and to calculate the average rate constant. Furthermore, the use of a pseudo-second-order reaction is no more than an approximation which would not be expected to apply in the general case or over a wide range of conditions.

In 1959, Andrews [28] obtained a series of empirical models that used the Voorhies relationships and appeared to fit adequately data from both fixed-bed and fluid crackers. All such models contain a number of inconvenient features which limit their use. For example, they seem to be overly simplified in the face of the very complex processes that occur in the cracking of gas oil. Furthermore, they require such measurements as the percentage of carbon on catalyst in order to calculate an empirical rate constant. Such measurements are very difficult to make in real time. The models are also quite narrow in the range of conditions they describe with one set of parameters.

A newer and more rigorous model was that proposed by Weekman [29]. His development begins with writing the continuity equations for a fixed-bed plug-flow reactor and assumes that feed space time is much shorter than catalyst time on stream. With these assumptions he obtains a differential equation relating the rate of conversion with the distance along the reactor bed. In order to take into account volume expansion that occurs on cracking, and the change in the refractoriness of the feed, Weekman assumes that second-order kinetics apply. The resultant equation is

$$\frac{dy}{dz} = \frac{F \rho_v}{\rho_0 S} k_0 (e^{-\lambda \theta}) y^2 \tag{6.10}$$

where y is the instantaneous fraction of gas oil converted, z the distance along the catalyst bed, F the void fraction in the bed, ρ_0 the density of gas oil, ρ_v the vapor density, k_0 the initial rate constant, θ the normalized time on stream, λ a deactivation parameter, and S the space velocity. By integrating Equation

(6.10), one obtains the instantaneous fraction of gas oil converted at a given time:

$$y = \frac{1}{1 + A_w Ze^{\lambda\Theta}} \tag{6.11}$$

where

$$A_w = \frac{F\rho_v k_0}{\rho_o S} \tag{6.12}$$

Finally, one can evaluate the average cumulative fraction converted by means of the equation

$$\overline{X} = \int_0^1 y \, dz = \frac{1}{\lambda} \ln \frac{1 + A_w}{1 + A_w e^{-\lambda\Theta}} \tag{6.13}$$

Subsequently, Weekman extended his model to describe the case of a moving-bed reactor:

$$X = \frac{A_w(1 - e^{-\lambda})}{1 + A_w(- e^{-\lambda})} \tag{6.14}$$

and for a fluidized-bed reactor,

$$X = \frac{A_w}{1 + \lambda + A_w} \tag{6.15}$$

Weekman's elegant mathematical model is attractive in its simplicity but suffers some of the limitations of its predecessors by assuming a second-order reaction to account for feed refractoriness, while assuming an exponential expression for catalyst decay as a function of time on stream.

In 1971, Pachovsky and Wojciechowski [30] published a model of gas-oil cracking based on the mechanism first proposed by Campbell and Wojciechowski [31]:

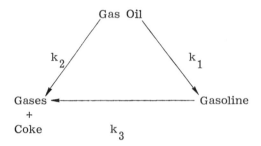

The model assumes a first-order rate expression for the conversion of each of the hydrocarbons found in the feed.

$$-\frac{dC_A}{d\tau} = (k_1 + k_2)C_A \qquad (6.16)$$

where τ is the space time in a fixed-bed reactor operating in plug flow, C_A the gas-oil concentration, and k_1 and k_2 the rate constants for the mechanism presented above. To account for catalyst decay, the rate constants k_1, k_2, and k_3 were modified using the time-on-stream function proposed by Wojciechowski [32]:

$$k = k_0(1 + Gt)^{-N} \qquad (6.17)$$

where G and N are deactivation parameters and t is the time on stream. The decay function arises from the assumption that catalyst deactivation is a function of active-site concentration and the time of exposure of the catalyst to reaction conditions; it was described in more detail in Chapter 4. The function used by Weekman [29] is a specific case of this general assumption and applies when decay is dependent on the first order of the active-site concentration.

To account for the refractoriness of the feed, the expression developed by Kemp and Wojciechowski [24] was used to modify the rate constant as a function of conversion. Thus the final rate constant as used in this approach takes the form

$$k = k_0 (1 + Gt)^{-N} \left(\frac{C_A}{C_{A0}}\right)^W \tag{6.18}$$

where C_{A0} is the initial concentration of gas oil and W is the refractoriness parameter and takes into account the increasing difficulty of cracking the residue as feed is converted during a reaction. Substituting the modified rate constant above in the rate expression permits one to calculate the instantaneous fraction converted, X_A, in a fixed-bed reactor using the following expression, which also allows explicitly for volume expansion:

$$(k_{10} + k_{20})(1 + Gt)^{-N} Pbt_f = \int_0^{X_A} \left(\frac{1 + \epsilon X_A}{1 - X_A}\right)^{1+W} dX_A \tag{6.19}$$

where b is a proportionality constant, P the catalyst-to-oil ratio, and ϵ the volume expansion due to cracking.

The instantaneous fraction converted can be used to calculate the average conversion at the end of a run by the expression

$$\overline{X} = \frac{1}{t_f} \int_0^{t_f} X_A \, dt \tag{6.20}$$

These equations have been used to describe catalytic cracking of various gas oils on a variety of catalysts and have proved to be quite adequate in fitting data over a wide range of conditions [2,33,34]. They have the advantage of allowing the decay expression to vary according to the specific behavior of each

feed/catalyst system, as well as allowing for the specific feed
interaction with the catalyst by means of the refractoriness
parameter W. Unfortunately, the necessary calculations are complex and cannot be done without the use of digital computers.
This not only complicates the matter of data interpretation but
also makes it difficult to develop an intuitive grasp of the behavior of the system.

Other cracking models have been proposed by various workers,
such as the elaboration by Prasad and Doraiswamy [35], who
extended Weekman's model for the general nth-order reaction, and
of Gustafson [36] and Pryor and Young [22], who assumed a
second-order reaction and first-order deactivation. The development of all modern cracking models [36] recognizes that the
presence of the various constituents in gas oil will have a major
influence on the form of the kinetic expressions that result.
Furthermore, it will influence the parameters obtained from the
models, since most models use lumped parameters and the nature
of the lumps used will influence the rate constants corresponding
to the lumps.

Aris and others [37,38] have pointed out that even if all the
reactions are of first order, the distribution of reactivities in the
initial composition of the feed will influence the apparent order of
the overall reaction. This is also the conclusion of the refractoriness development of Kemp and Wojciechowski [24]. Many other
authors have found that the lumping of a large number of parallel
first-order reactions with different rate constants results in an
apparent overall order greater than 1 [39,40]. The general order
W proposed by Kemp and Wojciechowski allows a quantitative comparison of various feedstocks by means of the single parameter
without decomposition of the feed into the various lumps or components. This presents an attractive method of treating data
over a wide range of conditions but does not allow predictive
statements to be made since W cannot be constructed from knowledge of the feed composition. What is needed is a model for predicting W from suitable lumps. Work such as that of Ozawa [41],
who has presented criteria for grouping various components of
the feed without first obtaining the complete set of corresponding
rate constants, may show the way. Such an approach may yet
lead to a quantitative formulation for W from knowledge of the
feed composition.

6.3.4 Gasoline Selectivity

It is clear that a mathematical description of gas-oil cracking which is capable of predicting only the conversion is of limited utility since it is the selectivity of cracking that is of principal interest in the commercial process. Because of this, models have been developed which, besides predicting conversion, are also capable of predicting selectivity, especially selectivity for gasoline. For instance, Weekman and Nace [9] extended the conversion model of Weekman to introduce gasoline selectivity. They lump their products into gas oil, gasoline, and gases in order to describe the system in kinetic terms. Subsequent work has shown that the rate constants in this simplified mechanism change with feed composition and can be correlated with the paraffin, naphthene, and aromatic content of the feed [19,20]. Such correlations are valid in the range of feed compositions studied but should not be extended beyond this range until such time as the correlations are better understood. Subsequently, Jacob et al. [10] presented a kinetic scheme which was independent of the initial composition of the feed. In this work they followed the initial assumptions made in developing Weekman's equations and arrived at the following equations:

$$\frac{dy_1}{dX} = \frac{\rho_v}{\rho_o S} k_0 \phi_0 y_1^2 \qquad (6.21)$$

$$\frac{dy_2}{dX} = \frac{\rho_v}{\rho_o S} (k_1 \phi_1 y_1^2 - k_2 \phi_2 y_2) \qquad (6.22)$$

where y_2 is the instantaneous fraction of gasoline produced, ϕ the deactivation function for the cracking to gasoline, and k_2 the corresponding rate constant at zero time. Quantities subscripted with a 1 refer to the formation of gases. Once again it was assumed that the deactivation functions, both ϕ_1 and ϕ_2, were approximated by the same exponential function of time on stream.

From this the following equation for the formation of gasoline was developed:

$$y_2 = r_1 r_2 e^{-r_2/y_1} \left[\frac{1}{r_2} e^{r_2} - \frac{y_1}{r_1} e^{-r_2/y_1} - E_{in}(r_2) + E_{in} \frac{r_2}{y_1} \right]$$

$$(6.23)$$

where

$$r_1 = \frac{k_1}{k_0} \qquad r_2 = \frac{k_2}{k_0} \qquad E_{in}(x) = \int_{\infty}^{X} \frac{e^x}{x} \, dx$$

From this, the average gasoline yield $\overline{y}_2$ can be computed:

$$\overline{y}_2 = \int_0^1 y_2 d\phi \qquad\qquad (6.24)$$

The value of $\overline{y}_2$ is calculated by numerical methods. These equations and their modifications for the moving- and fluidized-bed reactors have been fitted to experimental data with satisfactory results, although the methods used for data fitting and validation of the model were not statistically proper [42].

In developing their model, Pachovsky and Wojciechowski [30] applied the same triangular mechanism as that used by Weekman [9] and by Campbell and Wojciechowski [31] and based their treatment on the following continuity relationships:

$$\frac{\delta C_A}{dt} = U_v \frac{\delta C_A}{\delta z} + C_A \frac{\delta U_v}{\delta z} = -R_1(C_A, t) \bigg|_A \quad -R_2(C_A, t) \bigg|_A$$

$$(6.25)$$

$$\frac{\delta C_B}{dt} + U_v + C_B \frac{\delta U_v}{\delta Z} = R_1(C_A, t) \bigg|_B \quad -R_3(C_B, t) \bigg|_B \qquad (6.26)$$

$$\frac{\delta C_c}{dt} + U_v \frac{\delta C_c}{\delta z} + C_c \frac{\delta U_v}{\delta z} = R_2(C_A,t)\bigg|_C + R_3(C_B,t)\bigg|_C \qquad (6.27)$$

where C_A, C_B, and C_C are, respectively, the instantaneous concentrations of the original feed, gasoline, and $C_4(-)$ gases plus coke; τ is the instantaneous catalyst time on stream; z is the axial distance along the fixed-bed reactor; U_v is the vapor velocity in the z direction; and R_1, R_2, and R_3 are the rates of reaction with respect to the corresponding branches of the mechanism. By changing the axial direction z into a function of space time and accepting that the final time on stream is much larger than the space time, the equations above are reduced to a system of ordinary differential equations:

$$\frac{dC_A}{d\tau} + \frac{C_A}{U_v} \frac{dU_v}{d\tau} = -R_1(C_A,t)\bigg|_A \; -R_2(C_A,t)\bigg|_A \qquad (6.28)$$

$$\frac{dC_B}{d\tau} + \frac{C_B}{U_v} \frac{dU_v}{d\tau} = R_1(C_A,t)\bigg|_B \; -R_3(C_B,t)\bigg|_B \qquad (6.29)$$

$$\frac{dC_C}{d\tau} + \frac{C_c}{U_v} \frac{dU_v}{d\tau} = R_2(C_A,t)\bigg|_C \; + R_3(C_B,t)\bigg|_C \qquad (6.30)$$

Rate expressions for the various rates of reaction were formulated using the same method as described before.

$$R(C_A,t)\bigg|_A = k_0 \, C_S \, C_A \left[\frac{C_A}{C_{AO}}\right]^W \qquad (6.31)$$

$$= k_0 \, C_{SO} \, C_A \left[\frac{C_A}{C_{AO}}\right]^W (1 + Gt)^{-N} \qquad (6.32)$$

When this equation is substituted into the system of differential equations presented above and the concentrations are converted to the weight fractions, one obtains the following:

$$\frac{dX_A}{d\tau} = (k_{10} + k_{20}) \left[\frac{1 - X_A}{1 + \varepsilon X_A} \right]^{1+W} (1 + Gt)^{-N} \qquad (6.33)$$

$$\frac{dX_B}{d\tau} = \left| k_{10} \left[\frac{1 - X_A}{1 + \varepsilon X_A} \right] - k_{30} \left[\frac{1}{1 + \varepsilon X_A} \right] X_B \right| (1 + Gt)^{-N}$$

$$(6.34)$$

$$\frac{dX_C}{d\tau} = \left| k_{20} \left[\frac{1 - X_A}{1 + \varepsilon X_A} \right] + k_{30} \left[\frac{1}{1 + \varepsilon X_A} \right] X_B \right| (1 + Gt)^{-N}$$

$$(6.35)$$

The weight fractions of A, B, and C represent instantaneous values at a given contact time and at an instantaneous catalyst age t. Since instantaneous values are not normally measured in such systems, the time-averaged values $\overline{X}$ are calculated using

$$\overline{X}_i = \frac{1}{t_f} \int_0^{t_f} X_i \, dt \qquad (6.36)$$

In most cases it is convenient to use the catalyst-to-oil ratio as a variable rather than the contact time. The relationship

$$\tau = Pbt_f \qquad (6.37)$$

is therefore often introduced into numerical computations in this approach. Here P is the catalyst-to-oil weight ratio and b is a proportionality constant whose value can be computed from the relationship

$$b = \frac{\rho_v}{\rho_c} \tag{6.38}$$

where ρ_v is the density of the vapor in the reactor inlet and ρ_c is the density of the catalyst.

This model has been applied successfully to a number of cracking feeds [2,25,34,43,44] over a variety of catalysts and has yielded some interesting insights into the relationships that exist between the parameters obtained and the composition of the system under study [25]. Among the conclusions drawn are: that in a series of feeds consisting of a dewaxed and deasphalted neutral distillate combined with various amounts of wax, there is no evidence of interaction between these two lumps; that the coke made from both the waxy and the dewaxed feeds was a secondary product of the reactions; that the presence of wax increases the rate of catalyst decay; and that the addition of wax to a dewaxed feed increases the heterogeneity of the feeds, as evidenced by an increase in the W parameter.

All the foregoing observations are in agreement with the various points raised at the beginning of this chapter. It seems that the only new complication introduced by feed heterogeneity is the use of the refractoriness parameter, W.

6.4   CRACKING OF HEAVY FEEDSTOCKS

As the refining industry expands its demand for feedstocks and the pace of new oil discoveries slackens, attention turns to heavier cracking feedstocks. Not only is the heavier feed less expensive and more readily available, but its use in applications other than cracking is decreasing, while the demand for motor fuels and naphtha is expected to keep increasing in the long run [45].

The heavy feeds differ from the light feeds in their final boiling point and in composition [46]. The boiling point is generally so high that in the commercial reactor part of the feed contacts the catalyst as a fine liquid spray. This contacting method must have a substantial effect on the various diffusional problems mentioned in previous discussions. The composition is also such that fractions such as polycycloalkane, mono- and polynuclear aromatics, and asphaltenes are a significant part of the heavy feed [47]. These fractions are associated with heteroatom-containing molecules whose sulfur content can range up to about 6% while the nitrogen

content can reach about 0.7%. Combined with these high concentrations of heteroatom catalyst inhibitors are large amounts of
metals, reaching 2000 ppm in crudes being processed [48,49].

The poisonous nature of these feed components is combined
with a low hydrogen-to-carbon ratio in the feed itself, due to the
condensed polynuclear aromatics and other materials that contribute
to the Conradson carbon of the feed. The challenge is therefore
either to find catalysts that can successfully contend with such
feeds, or to treat the feed in some way so that it is more in line
with light feed properties.

Several options are available to remove or transform the undesirable components [50]:

1. Methods that produce two fractions from the heavy feed, one
   with a high, the other with a low H/C ratio [51,52]. Thermal
   coking is such a process. It also removes metals and poisoning heteroatoms. Since the process is thermal it lacks the
   desired selectivity of catalytic processes.

2. Feed pretreatments such as hydrotreating have the advantage
   of increasing the H/C ratio while reducing heteroatom and
   metals content [53].

3. Physical or chemical separation of heavy feed components by
   acid treatment or deasphalting. This removes some metals
   and heteroatoms and reduces Conradson carbon [54,55].

These and variations which fall under these headings all involve additional processing costs and installations. The trend has
therefore been to encourage the development of catalysts that can
withstand the demands of heavy oil cracking. Since the 1960s,
refiners have resorted to direct cracking, using a variety of approaches [56]. Some of these, such as an increase in O_2 partial
pressure in the regenerator, were meant to deal with the symptoms,
in this case the high coke make, rather than trying to solve the
cracking problems. These can best be tackled by arriving at a
suitable catalyst formulation.

From previous discussion of the cracking reaction one can propose the specifications for the heavy oil cracking catalyst [57—59].
To deal with the large molecules, one needs large pores. At the
same time, a zeolite component will help to improve product quality
by secondary reactions. The large pores should be on a matrix
of fair catalytic activity, since the liquid feed and the large molecules therein are unlikely to see the interior of any zeolite

present. By using a large-pore catalyst and adding a passivating agent to the feed, metals poisoning can be controlled. The hetero-atom inhibition is controlled by the zeolite content, as these mate-rials seem to be more tolerant of this kind of poison.

In this way we will have formulated a catalyst with large matrix pores, containing enough zeolite to handle the poisons, and we propose to introduce passivating agents to control metals activity. The next problem that requires attention is the high coke make.

In principle, the addition of suitable hydrogen transfer agents to the heavy oil feed will control coke make. Whether agents that are cheap enough can be found remains to be seen. The trend has been to designing catalysts which can deal with high values of coke on catalyst. The coke is usually burned off at conditions which are more severe than would be used for low coke yields. This necessitates a catalyst that is hydrothermally more stable than usual. This in turn requires the exchange of the sodium present in the catalyst by stabilizing cations to levels as low in sodium as possible. The large-pore structure of the matrix is also a help, as the catalyst is less likely to sinter in the regenerator.

The remaining problem is the fact that large coke make is con-nected with rapid deactivation. This means that high catalyst-to-oil ratios have to be used in order to achieve reasonable conver-sion, and hence catalyst circulation rates are high. This in turn requires a hard catalyst to withstand vigorous abrasion. The large-pore matrix is also compatible with this requirement.

In the regenerator, the additional gas flows required for re-generation can lead to large catalyst losses up the stack unless the cyclones are aided in their separation by being presented with a dense catalyst. The catalyst density should therefore be as high as possible, another feature that is compatible with a large-pore structure in the matrix.

The cracking of heavy feeds on the catalyst designed above will proceed by chemical and transport mechanisms which are poorly understood. The nature of the process is such that it is almost "catalytic coking." Nevertheless, the product distribution obtained by this process is substantially more attractive than that from thermal coking. Because of the low H/C ratio of the feed, little can be done to improve product yields unless a catalyst can be found which will be active and selective in hydrogen-transfer reactions. Such a development would present a clear example of a synergistic interaction in mixed feeds consisting of light and heavy oils, a possibility discussed in Section 5.7 and long sus-pected but never commercialized.

6.5   ROLE OF VARIOUS ACTIVE SITES IN
       GAS-OIL CRACKING

In considering the cracking of pure paraffins, we discussed the
various hypotheses as to the nature of the initiating event in the
cracking reaction. The complications are accentuated in the case
of gas-oil cracking, where one is dealing with a feed consisting
of a great variety of molecular species. As a result, most of the
discussion of gas-oil cracking behavior has concentrated on superficial phenomena such as total conversion or gross selectivity.
There is no doubt, however, that if the values of the kinetic
parameters involved in gas-oil cracking were established, a great
deal of information could be obtained by observing their behavior
with systematic changes in catalyst and feed composition. Corma
and Wojciechowski [43] have attempted to explain the involvement
of the various types of active sites at the different stages in the
catalytic cracking of gas oil. Comparing the kinetic parameters
obtained by fitting the time-on-stream model to data obtained by
cracking the same gas oil on two different zeolite catalysts, they
attempted to explain the influence of the various types of sites on
the catalytic cracking of this gas oil.
     Since the feed was identical in both cases, it is clear that any
differences observed in the parameters shown in Table 6.1 must
be related to the nature of the catalyst and therefore to the
nature of the active sites present on the catalyst. From information obtained by IR spectroscopy, and from cumene cracking used
as a test reaction, it is known that this particular HY zeolite has
more Brønsted sites and fewer Lewis sites than does LaY [58].
On the other hand, from studies of the acid strength distribution
as measured by Benesi's method, it was established that the number of active sites of $pK < 6.8$ is greater in HY, whereas LaY
has a larger number of strong acid sites with $pK < 1.5$ [43]. It
is the stronger sites that are believed to be active in the catalytic
cracking of paraffins [59]. In light of all these considerations,
one can form the following picture of the cracking of this particular paraffinic gas oil.
     The initiation reaction is mainly the abstraction of an $H^{\ominus}$ by
a strong Lewis acid on the catalyst surface. Thus k_0, the initial
rate constant for cracking, is expected to increase as the density
of Lewis sites is increased on the catalyst or as the feedstock is
made more susceptible to Lewis attack. If another feed were to
contain olefins or alkylated aromatics with side chains of C_2 or
larger, Brønsted sites would be expected to play a more prom-
inent role in the initiating reaction, leading to a different ratio

Table 6.1 Parameter Values for the Cracking of an Extracted and Dewaxed Neutral Distillate AlWO over HY and LaY Zeolite Catalysts

	482°C		503°C		524°C	
	HY	LaY	HY	LaY	HY	LaY
k_0 $(h^{-1}) \times 10^5$	2.92	76.2	5.16	195.0	10.08	472.1
G (h^{-1})	11.7	0.30	9.92	0.51	1.12	0.80
k_{md} (h^{-1})	12.45	9.01	18.04	10.4	9.33	15.6
N	1.06	31.0	1.83	29.7	8.13	24.7
m (order of decay)	1.94	1.03	1.55	1.03	1.12	1.06
W	1.08	3.08	1.25	3.24	2.50	3.34

Source: Ref. 43.

of k_0 for the two catalysts. On the other hand, the cracking
process takes place in the presence of olefins soon after the
initiating event occurs, and therefore the role of Brønsted acids
must be taken into account in the overall cracking reaction. This
effect is associated with refractoriness as discussed above and is
accounted for by the parameter W.

It should be made clear that Lewis acids and Brønsted acids
will not respond in the same way to the refractoriness of the feed.
From Table 6.1 we see that W is lower for the catalyst with more
Brønsted acid sites. This indicates that the Brønsted acid sites
present on these catalysts see the gas-oil feed as being more
homogeneous in reactivity than do the Lewis sites. At the same
time, the catalyst with the greater number of Brønsted acid sites
shows a higher order for the decay reaction and is subject to more
rapid decay.

Such considerations lead to the conclusion that the many-faceted
properties of cracking catalysts, all with their specific influence on
activity, selectivity, and catalyst decay, offer a fertile field for
optimization of catalyst/feed systems for the desired product distribution, reaction conditions, and reactor heat balances.

It is too soon to say whether studies of this type will lead to
useful ideas for the rational formulation of better catalysts. There
is a major problem with this approach in that even if the desired
catalyst properties could be rationally specified, it is not clear
that a suitable material can be tailored. Nevertheless, it is a
historical fact that science and technology advance on the basis
of a systematization and fundamental understanding of the subject
matter. Thus far, in catalytic cracking, these prerequisites have
not been adequately met.

REFERENCES

1. B. S. Greensfelder, H. H. Voge, and G. M. Good, *Ind. Eng.
    Chem.*, *41*: 2573 (1949)

2. A. Borodzinski, A. Corma, and B. W. Wojciechowski, *Can. J.
    Chem. Eng.*, *58*: 219 (1980)

3. P. B. Venuto and E. T. Habib, *Fluid Catalytic Cracking with
    Zeolite Catalysts*, New York, Marcel Dekker, 1979

4. C. J. Planck, D. J. Sibbett, and R. B. Smith, *Ind. Eng.
    Chem.*, *49*: 742 (1957)

5. B. C. Gates, J. R. Katzer, and G. C. A. Schmit, *Chemistry of Catalytic Processes*, New York, McGraw-Hill, 1979

6. T. M. John and B. W. Wojciechowski, *J. Catal.*, *37*: 240 (1975)

7. P. E. Eberly, Jr., C. N. Kimberlin, W. H. Miller, and H. V. Drushel, *Ind. Eng. Chem.*, *Process Des. Dev.*, *5*: 193 (1966)

8. D. E. Walsh and L. D. Rollman, *J. Catal.*, *49*: 369 (1977)

9. V. W. Weekman, Jr. and D. M. Nace, *AIChE J.*, *16*: 397 (1970)

10. S. M. Jacob, B. Gross, S. E. Voltz, and V. W. Weekman, Jr., *AIChE J.*, *16*: 701 (1976)

11. W. G. Appleby, J. W. Gibson, and G. M. Good, *Ind. Eng. Chem.*, *Process Des. Dev.*, *1*: 102 (1962)

12. R. W. Blue and C. J. Engle, *Ind. Eng. Chem.*, *43*: 494 (1951)

13. K. M. Watson and E. F. Nelson, *Ind. Eng. Chem.*, *25*: 880 (1933)

14. E. F. Nelson, *Oil Gas J.*, *50*(24): 145 (1951)

15. K. van Nes and H. A. van Westen, *Aspects of Constitution of Mineral Oils*, New York, Elsevier, 1951

16. H. E. Reif, R. F. Kress, and J. S. Smith, *Pet. Refiner*, *40*(5): 237 (1961)

17. M. E. Fitzgerald, J. L. Moirano, H. Morgan, and V. A. Cirillo, *Appl. Spectrosc.*, *24*: 106 (1970)

18. P. J. White, *Hydrocarbon Process.*, *47*(5): 103 (1968)

19. D. M. Nace, S. E. Voltz, and V. W. Weekman, Jr., *Ind. Eng. Chem.*, *Process Des. Dev.*, *10*: 530 (1971)

20. S. E. Voltz, D. M. Nace, and V. W. Weekman, Jr., *Ind. Eng. Chem.*, *Process Des. Dev.*, *10*: 538 (1971)

21. S. E. Voltz, D. M. Nace, S. M. Jacob, and V. W. Weekman, Jr., *Ind. Eng. Chem.*, *Process Des. Dev.*, *11*: 261 (1972)

22. J. N. Pryor and G. W. Young, *Stud. Surf. Sci. Catal.* (Catal. Energy) (Kaliaguine and Mahay, Eds.), Amsterdam, Elsevier, p. 173, 1984

23. H. Owen, P. W. Snyder, and P. B. Venuto, *Proc. 6th Int. Congr. Catal., London, 2:* 1071 (1976)

24. R. R. D. Kemp and B. W. Wojciechowski, *Ind. Eng. Chem., Fundam., 13:* 332 (1974)

25. R. A. Pachovsky and B. W. Wojciechowski, *Can. J. Chem. Eng., 53:* 308 (1975)

26. A. Voorhies, Jr., *Ind. Eng. Chem., 37:* 318 (1945)

27. F. H. Blanding, *Ind. Eng. Chem., 45:* 1186 (1953)

28. J. H. Andrews, *Ind. Eng. Chem., 51:* 507 (1959)

29. V. W. Weekman, Jr., *Ind. Eng. Chem., Process Des. Dev., 7:* 90 (1968)

30. R. A. Pachovsky and B. W. Wojciechowski, *Can. J. Chem. Eng., 49:* 365 (1971)

31. D. R. Campbell and B. W. Wojciechowski, *Can. J. Chem. Eng., 47:* 413 (1969)

32. B. W. Wojciechowski, *Can. J. Chem. Eng., 46:* 48 (1968)

33. R. A. Pachovsky and B. W. Wojciechowski, *Can. J. Chem. Eng., 50:* 306 (1972)

34. J.-J. Yeh and B. W. Wojciechowski, *Can. J. Chem. Eng., 56:* 599 (1978)

35. K. B. S. Prasad and L. K. Doraiswamy, *J. Catal., 32:* 384 (1974)

36. W. R. Gustafson, *Ind. Eng. Chem., Process Des. Dev., 11:* 507 (1972)

37. R. Aris and G. R. Gavalas, *Philos. Trans. R. Soc. London, A260,* 351 (1966)

38. R. Aris, *Arch. Ration. Mech. Anal., 27:* 356 (1968)

39. D. Luss and P. Hutchinson, *Chem. Eng. J., 1:* 129 (1971)

40. D. Luss and P. Hutchinson, *Chem. Eng. J., 2:* 172 (1971)

41. Y. Ozawa, *Ind. Eng. Chem., Fundam., 12:* 191 (1973)

42. G. E. P. Box and N. R. Draper, *Biometrika, 52:* 355 (1965)

43. A. Corma and B. W. Wojciechowski, *Can. J. Chem. Eng., 60:* 11 (1982)

44. A. Corma, J. Juan, J. Martos, and J. Molina, *Proc. 8th Int. Congr. Catal.*, *Berlin*, 2: 293 (1984)

45. C. P. Carter, *NPRA Ann. Meeting*, AM82-38, 1982

46. J. S. Ball, W. E. Haines, and R. V. Helm, *Proc. 5th World Pet. Congr.*, 5: 175 (1959)

47. M. A. Bestougeff, *Fundamental Aspects of Petroleum Geochemistry* (B. Nagy and U. Colombo, Eds.), p. 77, Amsterdam, Elsevier, 1967

48. P. B. Venuto and E. T. Habib, *Catal. Rev. Sci. Eng.*, 18: 1 (1978)

49. Anon., *Oil Gas J.*, 73(21): 94 (1975)

50. A. G. Bridge, G. D. Gould, and J. F. Berkman, *Oil Gas J.*, 79(3): 85 (1981)

51. A. M. Edelman, C. R. Lipuma, and F. G. Turpin, *Proc. 10th World Pet. Congr.*, 10(4): 167 (1980)

52. F. N. Dawson, *Hydrocarbon Process.*, 60(5): 86 (1981)

53. R. E. Ritter, W. A. Welsch, L. Rheaume, and J. S. Magee, *NPRA Ann. Meeting*, AM-81-44, 1981

54. C. K. Viland, *Pet. Refiner*, 37(3): 197 (1958)

55. J. A. Gearhart, *NPRA Ann. Meeting*, AM-80-34, 1980

56. G. P. Masologites and L. H. Beckberger, *Oil Gas J.*, 71(48): 49 (1973); D. Decroocq, *Catalytic Cracking of Heavy Petroleum Fractions*, Paris, Technip Editions, 1984

57. J. S. Magee, R. E. Ritter, and L. Rheaume, *Hydrocarbon Process.*, 58(9): 123 (1979)

58. J. M. Maselli and A. W. Peters, *Catal. Rev. Sci. Eng.*, 26: 525 (1984)

59. R. E. Ritter, *Natl. Pet. Refiners Assoc.*, AM-84-57, 1984

7

The State of the Art and
the Beginnings of a Science

The field of catalytic cracking is full of superlatives. Cracking
uses the largest tonnage of commercial catalysts. It produces the
largest volume of products. It has been instrumental in creating
the most varied, richest, and most influential of our traditional
industries. It has been the inspiration of more papers on kinetics,
product distribution, catalyst formulation, and other such topics
than any other single process in industry. It is therfore not surprising that a great deal is known about cracking catalysts and
catalytic cracking. What is surprising is that a coherent picture
of the chemistry of this process is only now beginning to emerge.
In brief, the picture is as follows.

Catalytic cracking occurs on acid catalysts containing strong
Brønsted and Lewis sites. It is accompanied by a vast variety of
other processes, such as skeletal isomerization, cyclization,
aromatization, disproportionation, dehydrogenation, and others.
Each of these processes occurs on sites of appropriate strength
and nature and not on other sites that may be present. The
weak sites catalyze reactions such as cis-trans isomerization,
stronger sites catalyze double-bond isomerization, followed in turn
by skeletal isomerization, cracking, and coke formation on sites
of increasing strength.

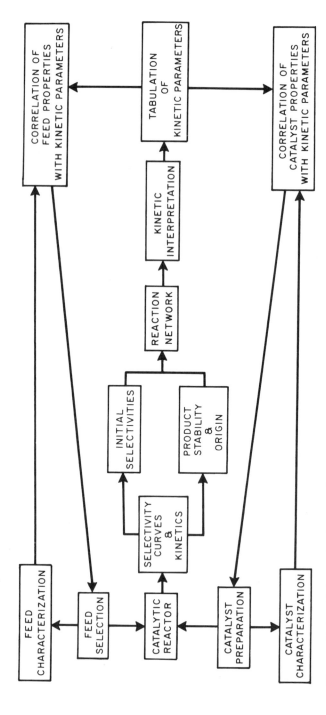

Figure 7.1   Schematic description of a rational procedure for cracking catalyst development.

228

Further complications are introduced by the size, tortuosity, and surface configuration of the pores present in crystalline cracking catalysts. Appreciation of the significance of these effects is new to the field of heterogeneous catalysis. It now seems certain that they have enormous potential to affect selectivity and to introduce specificity to the list of important properties of heterogeneous commercial catalysts.

Because of the great complexity and variety of commercial cracking feeds, the relative importance of the various processes accompanying cracking varies with the catalyst/feed combination. This leads to a vast range of possibilities in selecting the catalyst/feed combination in order to optimize the economics of a process. It is therefore of considerable importance that cracking catalyst formulation be guided by a rational, quantitative method of evaluation of the kinetics and selectivity of each such combination.

All hope has been lost that a simple test reaction such as cumene cracking will allow such an evaluation to be made. It is still possible to imagine that an appropriate selection of 5 or 10 test reactions using model compounds may adequately characterize a catalyst for industrial applications. An even more promising prospect of success lies in using 5 or 10 test reactions based on "lumped feeds," each composed of a range of pure paraffins or a range of alkylated aromatics or some other appropriate lump. Nevertheless, the most certain method of catalyst evaluation remains the use of complete commercial feeds. Significant progress has been made in the quantification of catalyst activity, selectivity, decay, and feed refractoriness for such feeds, in terms of meaningful parameters described here and elsewhere. Efforts can now be initiated to assemble such parameters and correlate them with feed composition. Such an undertaking, although large, pales in comparison with the quantity of less organized data already existing and constantly being produced by the various laboratories.

All the work of those who have illuminated this topic over the past four decades will have gone to waste unless we use their insights to develop a rational procedure for the development and evaluation of cracking catalysts. To this end, we suggest a sequence of operations such as that shown in Figure 7.1. Such an approach offers an opportunity to develop commercial cracking catalysts in a rational way and, at the same time, to generate a body of consistent, quantitative, information in this complex and important area of catalysis.

INDEX